Air Lasers: Challenge & Spectroscopy

Air Lasers: Challenge & Spectroscopy

Jacob

First Printing, 2024

Horace, my friend, a curious path
we've walked

Table of Contents

Chapter 1

Overview

You are most likely surrounded by air. This is easy to forget because air is abundant on earth and usually invisible. Air is an essential component of our existence because it allows the alveoli in our lungs to exchange carbon dioxide in blood for oxygen. We are a technology that relies on air. There must be air anywhere we go, so it is convenient to develop other technologies that only rely on air. This book explores air as the gain medium of a laser, which should allow us to remotely place a laser anywhere that we can breathe.

The "air laser" primarily refers to a gain medium that starts with only the atoms or molecules of air and an external pump laser. Sometimes, it refers to a hypothetical device that would use the gain medium to produce laser pulses. Ideally, the pump laser could initiate the air laser device at standoff distances, and it would produce a laser pulse that propagates backwards towards the source. This hypothetical air laser could be initialized in any direction, which enables remote spectroscopy over the entire field of view from a single location.

The vision of air lasing presents significant scientific and technical challenges. Focusing a powerful laser in air leads to nonlinear effects like self focusing, plasma defocusing, and self-phase modulation that distort the pulse during propagation. Gain is available in the plasma on transitions involving different vibrational states of the $B^2\Sigma_u^+$ and $X^2\Sigma_g^+$ electronic states of N_2^+ [2], but the excitation of gain dynamics is entwined with ionization and laser propagation. This generates a lot of interesting physics resulting in many different explanations for the lasing emission.

As the title of this book suggests, we attempt to simplify the experiment conditions to isolate the gain. First, we isolate gain from the nonlinear effects of propagation during filamentation by limiting the propagation distance, then we try to isolate the excitation of gain from ionization. You are reading the first brief chapter that provides an overview of the book. Chapter 2 follows, and it introduces the propagation of a laser pulse and air lasing during filamentation. The remainder of the book explores the behaviour of the N_2^+ gain medium.

Chapter 3 introduces our unique experimental approach that we utilize in the following chapters. To isolate gain, we use a narrow gas jet in vacuum to limit the propagation distance in the gain medium and prevent the accumulation of nonlinear effects. Like others, we use a separate probe pulse to measure gain. Chapter 3 discusses the optics and equipment inside and outside of the vacuum chamber that enable this measurement. It outlines the procedure to calibrate important experimental parameters like the laser intensity and polarization. It also demonstrates the analysis methods that produce the data presented in the figures through this book. Most of the work in this book formed the contents of three papers [, ,]. While Ch. 4, 5, and 6 are each based on one paper in chronological order, they build on each other as part of a larger story and introduce additional details.

We begin by investigating the source of gain itself in Ch. 4. The large excitation cross section for inelastic scattering of electrons with the $B^2\Sigma_u^+$ and $X^2\Sigma_g^+$ states suggests that excitation to the $B^2\Sigma_u^+$ state due to electron recollision during ionization could contribute to inversion [ ]. We halt recollision by increasing the ellipticity of the polarization of the pump pulse, but we only observe a small effect on gain. The results of experiments and a semiclassical calculation suggest that recollision can excite a few percent of the population, but it is not an essential mechanism. By comparing to measurements in air and using lower intensities close to the clamped intensity in a filament, we verify that this conclusion holds even in air filaments. As we vary the probe delay, these measurements reveal additional gain dynamics. Notably, the gain decays on two coexisting timescales that originate from either population decay or wave mixing.

These dynamics further reveal the source and behaviour of gain in Ch 5. We separate short- and long-term gain using different experimental conditions. The short-term gain is consistent with Raman gain in a V-system involving the $X^2\Sigma_g^+$, $A^2\Pi_u$, and $B^2\Sigma_u^+$ states, which appears at low pump intensity and does not require inversion between the $X^2\Sigma_g^+$ and $B^2\Sigma_u^+$ states. The long-term gain is consistent with an exponential-like decay of population inversion, and the large cross section for inelastic scattering of electrons also plays a role here. The rate of the exponential-like decay matches the rate of inelastic collisions between electrons and ions in the plasma that equalize the populations of the $B^2\Sigma_u^+$ and $X^2\Sigma_g^+$ states. As the populations equalize, any inversion decays. Superimposed on this exponential-like decay, molecular alignment of the lasing states imprints rotational dynamics on the gain. We observe a stronger rotational wave packet in the $B^2\Sigma_u^+$ state. The rotational wave packets decay on the same timescale as the decay of gain, which is consistent with the incoherent mixing of the rotational wavepackets in both states by the same inelastic collisions that equalize the populations.

The pump pulse interaction starts in neutral nitrogen before ionization, so the population dynamics and rotational wave packets are complicated by excitation that occurs before, during, and after ionization. Chapter 6 shows that the excitation by an additional non-ionizing pulse is more straightforward because it only interacts with the ion, and it enables control of the gain characteristics. This transforms the experiment to excite-probe spectroscopy on the N_2 ion itself. We show that the additional pulse annihilates or manipulates the original rotational wave packets with new conventionally generated wave packets. It also modifies the amount of gain in a straightforward way, which quenches or redirects the narrow-bandwidth emission following the probe pulse. These interactions highlight the pump- and control-induced coupling of the ground $X^2\Sigma_g^+$ state with a middle $A^2\Pi_u$ state, which is essential to gain.

Finally, Chapter 7 summarizes the main conclusions and discusses our perspective on the future of gain in N_2^+.

Chapter 2

Background

This chapter introduces key concepts that provide a foundation to discuss the gain in the nitrogen molecular cation. The final section of the chapter discusses the history of gain in N_2^+, which is a promising gain medium for the "air laser". To create an air laser pulse, an external laser pulse must first generate the ionized N_2^+ gain medium. This process is complicated in a couple of ways. First, the recipe for gain is mysterious and seems to require several independent ingredients. While it is relatively easy experimentally to produce gain, there is no complete theoretical understanding of how gain develops. Second, the pulse that generates the gain medium experiences nonlinear and linear effects during propagation in air, so the gain medium is inhomogeneous and uncontrolled as a function of propagation distance. The next chapter introduces a more controlled measurement to better-inform our understanding of gain.

2.1 A Laser Pulse

In most air lasing studies, one laser pulse creates the N_2^+ gain medium, a different laser pulse measures gain, and the resulting emission is an air laser pulse. The characteristics of these pulses are important to study, understand, and manipulate the gain in N_2^+. For example, a Gaussian function is a simple and convenient pulse shape for the purpose of calculations and illustrations.

2.1.1 Gaussian pulse

The envelope of the electric field of a Gaussian pulse

$$E(t) = \mathrm{Re}\left(E_0 \mathrm{e}^{i\omega_0 t} \mathrm{e}^{-\Gamma t^2}\right) \tag{2.1}$$

is a Gaussian function, where E_0 is the peak electric field amplitude, ω_0 is the central frequency of the pulse, and Γ determines the duration and chirp of the pulse. The real part of Γ determines the transform-limited pulse duration, while the imaginary part of Γ determines the chirp and corresponding instantaneous frequency. The instantaneous frequency

$$\omega_{inst}(t) = \omega_0 - 2t\,\mathrm{Im}(\Gamma) \tag{2.2}$$

is the time derivative of the phase [7].

The finite envelope of the laser pulse in time requires a distribution of frequencies. The pulse in time is related to the spectrum in frequency via a Fourier transform. For example, the Gaussian envelope in time yields a Gaussian distribution in the frequency domain

$$E(\omega) = E_0\sqrt{\frac{\pi}{\Gamma}} \cdot \mathrm{e}^{\frac{(\omega-\omega_0)^2}{4\Gamma}}, \tag{2.3}$$

where $E(\omega)$ is the Fourier conjugate of $E(t)$.

This analytical example illustrates the connection between time and frequency, but the amplitude and phase of the spectrum determine the shape of any pulse in time.

2.1.2 Electric field polarization

Equation 2.1 does not consider space. If the laser pulse is propagating in a direction $\vec{z}$, then the electric field oscillates in the plane normal to $\vec{z}$. In a simple and common example, the electric field is linearly polarized in one direction [*e.g.* $\vec{E}(t) = E(t)\vec{y}$].

Introducing a second component rotates the direction of the linear polarization assuming that both components oscillate in-phase. If the second component includes a π phase shift, the electric field amplitude never decreases all the way to zero and the electric field vector rotates. In that case, the polarization of the laser

$$\vec{E}(t) = E(t)\vec{y} + i\epsilon E(t)\vec{x} \tag{2.4}$$

can be quantified using the ellipticity ϵ. An ellipticity of zero corresponds to linear polarization, and an ellipticity of one corresponds to circular polarization. An elliptically polarized pulse can deliver a higher average intensity without a higher peak electric field.

This book only considers the case where the ellipticity is uniform across the transverse spatial profile of the beam, but the components in Eq. 2.4 can vary in the transverse plane in general. In our case, we can consider the polarization separately from the spatial profile.

2.1.3 Transverse spatial profile

A simple and common spatial profile is a Gaussian beam. A Gaussian beam refers to the fundamental Gaussian TEM_{00} mode of a cylindrically symmetric laser, which is a solution to the Helmholtz equation in the paraxial approximation [ ]. The spatial profile of a Gaussian beam in polar coordinates

$$E(r,z) = E_0 \frac{w_0}{w(z)} \exp\left\{\frac{-r^2}{w(z)^2}\right\} \exp\left\{-ik\frac{r^2}{2R(z)}\right\} \exp\{i\Phi(z)\} \tag{2.5}$$

depends on the beam radius

$$w(z) = w_0 \sqrt{1 + \left(\frac{z}{z_R}\right)^2}, \tag{2.6}$$

the radius of curvature of the phase fronts

$$\frac{1}{R(z)} = \frac{z}{z^2 + z_R^2}, \tag{2.7}$$

the Guoy phase $\Phi(z) = \mathrm{atan}\left(\frac{z}{z_R}\right)$, and the Rayleigh length $z_R = \frac{\pi w_0^2 n}{\lambda}$ [ ]. The propagation of this mode is determined by the radius of the beam waist w_0 and the wavelength λ. The transverse spatial profile is a Gaussian function at any point along z.

2.1.4 Initial laser source

Ultrafast lasers that produce near-infrared laser pulses with peak powers higher than 200 GW and durations shorter than 100 fs are widely available. The N_2^+ laser was initially discovered experimentally by using an ultrafast laser to ionize air [ ]. These lasers are ideal for studying the N_2^+ laser. High peak power ionizes more N_2 and creates a larger N_2^+ lasing medium [ ], and laser pulses with shorter duration resolve faster dynamics in N_2^+ [ ].

We use a commercial laser system based on titanium-doped sapphire to experimentally initiate and investigate the N_2^+ laser. The system produces 10 mJ linearly polarized laser pulses at a repetition rate of 1 kHz. The laser pulses are centered around 800 nm and have a minimum full width at half maximum (FWHM) pulse duration of 32 fs. The laser is split into two outputs to provide 5.5 mJ and 4.5 mJ sources with independent chirp control using separate grating compressors. These are relatively common input parameters for air lasing measurements.

2.2 Nonlinear Optics

Chapter 3 introduces an experimental approach that minimizes nonlinear effects on the pump laser pulse during the generation of the gain medium. In many experiments, the focused pump pulse propagates over long distances while creating the N_2^+ gain medium, and nonlinear effects modify it. Therefore, an initially simple laser pulse can develop elaborate structure during propagation.

Nonlinear optics refers to the nonlinear response of the material to an applied field. The material response can be represented as the polarization $P(t) = \epsilon_0 \chi E(t)$, which is the dipole moment per unit volume. The optical susceptibility χ can be expanded as a power series in the electric field $E(t)$

$$P(t) = \epsilon_0 \left(\chi^{(1)} + \chi^{(2)} E(t) + \chi^{(3)} E^2(t) + \ldots \right) E(t), \tag{2.8}$$

which adds higher order terms to the usual linear response $\chi^{(1)}$ [11].

These higher order terms can produce new frequencies in wave-mixing processes like harmonic generation, sum and difference frequency generation, and four wave mixing. For example, the second order susceptibility $\chi^{(2)}$ describes second harmonic generation. For a laser $E(t) \propto \mathrm{Re}\left(e^{i\omega_0 t}\right)$ with frequency ω_0, the nonlinear polarization due to the second order susceptibility

$$\begin{aligned} P^{(2)}(t) &= \epsilon_0 \chi^{(2)} E^2(t) \\ &\propto \epsilon_0 \chi^{(2)} \,\mathrm{Re}\left(e^{i2\omega_0 t}\right) \end{aligned} \tag{2.9}$$

oscillates at twice the frequency ($2\omega_0$). While the second order susceptibility is usually small (on the order of 10^{-12} m/V), this can be an efficient process using the high intensity of ultrashort pulses.

The third order susceptibility $\chi^{(3)}$ is of particular interest because it is nonzero in both centrosymmetric and non-centrosymmetric materials, whereas $\chi^{(2)}$ is zero in centrosymmetric materials. In addition to third harmonic generation and four wave mixing, $\chi^{(3)}$ is also responsible for the Kerr effect.

The refractive index including the third order susceptibility

$$\begin{aligned} n &= (1 + \chi)^{\frac{1}{2}} \\ &\approx 1 + \frac{1}{2}\chi \\ &= 1 + \frac{1}{2}\chi^{(1)} + \frac{1}{2}\chi^{(3)} E^2(t) \end{aligned} \tag{2.10}$$

depends on the intensity as

$$n = n_0 + n_2 I(t), \tag{2.11}$$

where n_0 is the usual linear refractive index, $n_2 \propto \chi^{(3)}$ is the nonlinear refractive index, and $I(t) \propto E^2(t)$ is the intensity of the laser. The Kerr effect has a number of consequences.

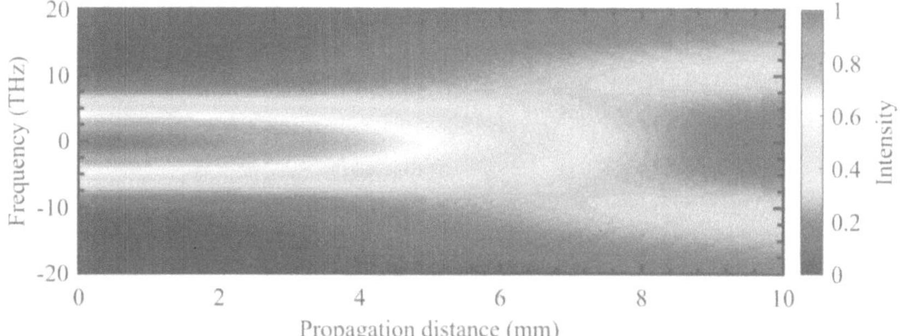

Figure 2.1: The spectrum of an initially Gaussian laser pulse centered at 800 nm with a duration of 30 fs and peak intensity of 5×10^{14} W/cm^2 in atmospheric nitrogen under the influence of self-phase modulation during propagation over 10 mm. $\Delta\Phi_{max} = 1.25\pi$ at 10 mm. Results calculated using the methods in Sec. 5.3.3.

2.2.1 Self-phase modulation

The phase of a pulse

$$E(z,t) = Re\left\{A(z,t)e^{-i(kz-\omega_0 t)}\right\} \tag{2.12}$$

depends on the index of refraction through the wave vector $k = n\omega_0/c$, where $A(z,t)$ is the envelope of the pulse and ω_0 is the center frequency. The total phase

$$\begin{aligned} \Phi &= \omega_0 t - kz \\ &= \omega_0 t - \frac{n_0\omega_0 z}{c} - \frac{n_2 I(t)\omega_0 z}{c} \end{aligned} \tag{2.13}$$

includes a contribution from the nonlinear refractive index, and the accumulated nonlinear phase

$$\Delta\Phi(t,z) = -\frac{n_2 I(t)\omega_0 z}{c} \tag{2.14}$$

is time-dependent due to the pulse shape in time [ ]. The instantaneous frequency

$$\omega_{inst} = \omega_0 - \frac{n_2\omega_0 z}{c}\frac{\mathrm{d}I}{\mathrm{d}t} \tag{2.15}$$

shows that new frequencies appear in the spectrum from self-phase modulation by the time dependent intensity profile. Redder and bluer frequencies appear on the rising and falling edge of the pulse, respectively.

Figure 2.1 is the spectrum of an intense short pulse under the influence of self-phase modulation as a function of propagation distance. After propagation, the spectrum broadened and developed two peaks and a trough. The phase accumulated over a propagation distance of L at the peak of the pulse

$$\Delta\Phi_{max} = \frac{n_2 I_0 \omega_0 L}{c} \tag{2.16}$$

is approximately equal to π multiplied by the number of troughs in the spectrum, where I_0 is the intensity at the peak of the pulse. Equation 2.16 equals π at 8 mm in Fig. 2.1.

2.2.2 Self-steepening

Due to the Kerr effect, self-steepening of the pulse in time occurs that accompanies the appearance of new frequencies in a spectrum. The optical path length

$$\begin{aligned} OPL &= nL \\ &= n_0 L + n_2 I_0 L \end{aligned} \tag{2.17}$$

is the product of the refractive index in the material and the geometric propagation distance, which increases with intensity due to the Kerr effect. The optical path length represents the equivalent propagation distance in vacuum, so the intense peak of a pulse propagates over an effectively longer distance than the wings. Therefore, the peak of the pulse is delayed in time relative to the wings of the pulse. This shortening of the pulse in time corresponds to a broader spectrum and creates an optical shock on the tailing edge [12]. This behaviour is not captured by Fig. 2.1.

2.2.3 Self-focusing

The Kerr effect also influences the spatial profile of a laser during propagation. Figure 2.2(a) shows the Kerr effect as a function of radial coordinate for a Gaussian beam. In this case, the optical path length is longest in the center and tapers off in the radial direction, which resembles a focusing lens. Figure 2.2(b) shows that two rays converge in the presence of this index of refraction. In other words, the spatially varying Kerr effect due to the transverse spatial profile of the laser pulse influences the divergence of the beam. While self-focusing always occurs, it must overcome diffraction to focus the beam.

Self-focusing overcomes diffraction at the critical power P_{cr}. The Kerr effect is significant when the accumulated nonlinear phase in Equation 2.16 equals roughly one

$$\frac{n_2 I_0 \omega_0 L}{c} \approx 1. \tag{2.18}$$

The accumulated phase depends on the propagation length L. For self-focusing to overcome diffraction of a Gaussian beam, the Kerr effect should be significant over the Rayleigh length ($L = z_R$). This requirement yields the critical power

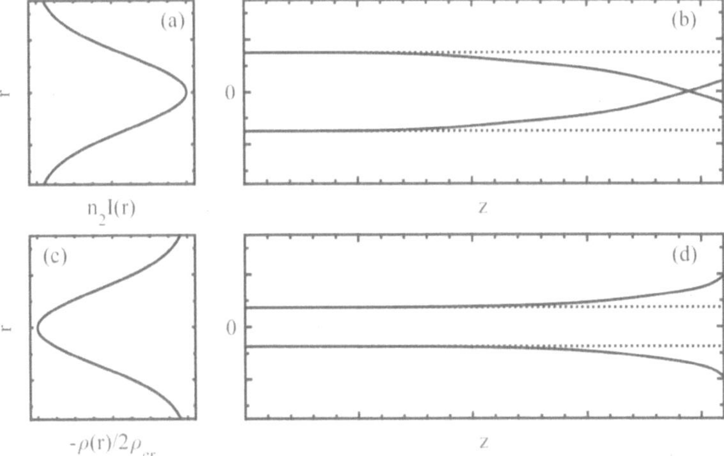

Figure 2.2: (a) The nonlinear contribution to the refractive index for a Gaussian beam. (b) The propagation of two rays with (solid) and without (dashed) the Kerr effect. (c) The reduced index of refraction due to a Gaussian density distribution of plasma. (d) Two rays with and without plasma defocusing. The rays were propagated using ray transfer matrix analysis.

$$\begin{aligned} \frac{2n_0n_2\pi^2w_0^2I_0}{\lambda^2} &\approx 1 \\ \implies P_{cr} &\approx \frac{\lambda^2}{2\pi n_0 n_2}. \end{aligned} \tag{2.19}$$

determined by the nonlinear refractive index and the wavelength, where $P_{cr} = \pi w_0^2 I_0$.

For a specific example, the critical power at $\lambda = 400\,\mathrm{nm}$ and $n_2 = 1 \times 10^{-19}\,\mathrm{cm^2\,W^{-1}}$ (for 100 kPa of N_2) is $P_{cr} = 2.6 \times 10^9\,\mathrm{W}$. This corresponds to a 100 µJ pulse with a duration of roughly 50 fs.

Plasma defocusing

When the laser ionizes the medium and generates plasma, the plasma introduces a negative contribution to the refractive index. Due to the intensity dependence of ionization, the plasma density also varies in the plane perpendicular to the propagation direction. In a simple case, it is most dense in the center and tapers off in the radial direction. Figure 2.2(c) shows the index of refraction for a Gaussian radial density distribution. This case is the opposite of self-focusing, and the equivalent rays in Fig. 2.2(d) diverge. Plasma can defocus the beam.

2.3 Gain and Absorption from Electronic Transitions

Gain appears on transitions in N_2^+. Population inversion can produce gain, but so can nonlinear wave mixing. Both types of gain are introduced here and discussed throughout this book.

2.3.1 Gain in an Inverted System

A two-level system provides a simple model of gain and absorption that is due to the relative population of states. While an inverted system usually provides gain as follows, absorption and gain only differ by a π phase shift.

The Hamiltonian operator for a simple two-level system is

$$H = \epsilon - \mu E. \tag{2.20}$$

Inserting the completeness relation on each side produces the density matrix representation

$$H = \begin{pmatrix} 0 & -\mu E \\ -\mu E & \epsilon \end{pmatrix}, \tag{2.21}$$

where the two levels in the system are separated in energy by ϵ, the electric dipole transition is allowed with transition dipole moment $\vec{\mu}$, and the electric field $\vec{E}$ is aligned with the transition dipole moment ($\vec{\mu} \cdot \vec{E} = \mu E$). The off diagonal elements are related to the Rabi frequency

$$\Omega_{10} = \frac{\mu E}{\hbar}. \tag{2.22}$$

The Liouville–von Neumann equation $\hbar\dot{\rho} = [H, \rho]$ determines the evolution of the density matrix in time from the initial condition, which yields coupled equations of motion

$$\begin{pmatrix} \dot{\rho}_{00} & \dot{\rho}_{01} \\ \dot{\rho}_{10} & \dot{\rho}_{11} \end{pmatrix} = -\frac{i}{\hbar} \begin{pmatrix} -\mu E(\rho_{10} - \rho_{01}) & \epsilon\rho_{01} - \mu E(\rho_{11} - \rho_{00}) \\ \epsilon\rho_{10} - \mu E(\rho_{00} - \rho_{11}) & -\mu E(\rho_{01} - \rho_{10}) \end{pmatrix}. \tag{2.23}$$

The on-diagonal density matrix elements are the state populations and the off-diagonal elements are the coherences between the states that represent the material response.

If a system of gaseous particles begins in thermal equilibrium, the population is distributed according to the Maxwell-Boltzmann distribution for the temperature of the particles. External processes can modify the state populations and coherences, but the excited population and coherence must decay as the system returns to equilibrium. This can be modeled as an exponential decay of population and coherence and included in Eq. 2.23 (*e.g.* see Sec. 6.2). The population decay T_1 is typically much longer than the coherence decay T_2. An external process like a preceding pump pulse can prepare the initial population, coherence, or both.

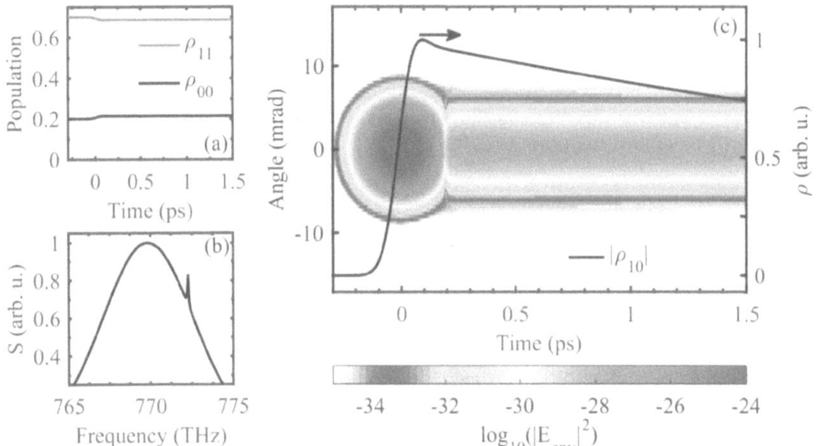

Figure 2.3: (a) The time evolution of the diagonal density matrix elements for initial conditions $\rho_{00}(t=0) = 0.2$ and $\rho_{11}(t=0) = 0.7$. The duration of the Gaussian pulse at $t = 0$ is 100 fs. The peak intensity is 10^8 W/cm^2, the transition dipole moment is 0.74 atomic unit (a.u.), and $\epsilon = 3.17$ eV. The coherence decay time is $T_2 = 5$ ps. (b) The laser spectrum after propagation of a single 0.8 μm step through a plane of atoms with density 1×10^{18} cm^{-3}. (c) The coherence (right axis) superimposed on the spatial profile of the pulse in the far field after propagation (color: logarithmic scale). The beam waist of the initially Gaussian beam in the near field is 50 μm.

If the populations of the two states are different, then Eq. 2.23 shows that coherence evolves while the field is on. When the field is on in the presence of coherence, it modifies the populations of the two states. This interaction produces Rabi oscillations of the populations as a function of time during the pulse [7, 11, 16]. For example, a relatively weak pulse induces a small fraction of a Rabi oscillation in Fig. 2.3(a). This is like a probe pulse weakly measuring an inverted system.

The polarization $\vec{P}$ in the wave equation

$$\nabla^2 \vec{E} - \frac{1}{c^2}\frac{\partial^2 \vec{E}}{\partial t^2} = \frac{1}{\epsilon_0 c^2}\frac{\partial^2 \vec{P}}{\partial t^2}, \tag{2.24}$$

is the dipole moment per unit volume. The polarization due to the transition

$$\begin{aligned} P_\mu &= N < \mu > \\ &= N\,\mathrm{Tr}(\rho\mu) \end{aligned} \tag{2.25}$$

connects the coherent material response back to the laser pulse, where N is the number density.

The decaying coherence generates emission that trails the pulse after propagation, which is known as free induction decay. The coherence time determines the duration and the linewidth of the stimulated emission. If the coherence decay is longer than the laser pulse, the spectral feature is narrow compared to the laser bandwidth. This can be the case for an ultrashort laser pulse amplified or absorbed by an electronic transition. Figure 2.3 shows these conditions.

The interaction in Figure 2.3(a) amplifies the pulse. Figure 2.3(b) shows the spectrum of the pulse after the interaction, which includes a narrow peak at the transition frequency on top of the initially smooth spectrum. Figure 2.3(c) shows the coherence superimposed on the spatial profile in the far-field. The spatial profile of the pulse in the near field is initially Gaussian, and the far-field is related to the near-field via Fourier transform [ ]. The figure only includes one propagation step, but the free induction decay grows more intense and complex during propagation. On subsequent propagation steps, the free induction decay itself is part of the interaction (*i.e.* μE).

Section 6.2 extends this development to three states in a V-configuration and discusses methods that provide the numerical solution to those equations.

2.3.2 Stimulated Raman scattering

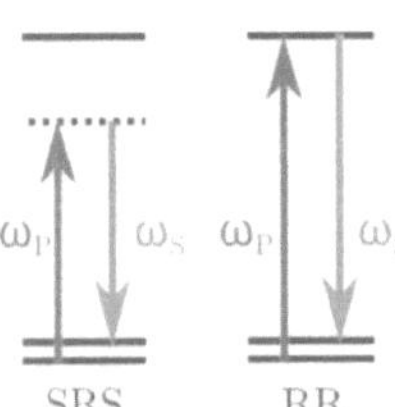

Figure 2.4: Stimulated Raman scattering (SRS) moves population between different states (solid lines) using a virtual state (dashed line), while a resonant transition enhances resonance Raman scattering (RR). In both cases, light is converted from ω_P to ω_S through a Stokes transition and the material is left in an excited state. The reverse process (anti-Stokes) is also possible.

Stimulated Raman scattering (SRS) is frequency-conversion and material excitation through simultaneous absorption and emission [ ]. SRS can also be a source of gain from electronic transitions. It is a nonlinear process that depends on the third order susceptibility $\chi^{(3)}$, so it can be classified as nonlinear optics (Section 2.2).

In ordinary Raman scattering, a pump photon (ω_P) excites the material to a virtual state that can spontaneously emit a signal photon (ω_S) and transition to a new state. In SRS, the emission of the signal photon is stimulated using laser light. This is illustrated in Figure 2.4 for a Stokes transition ($\omega_S < \omega_P$). SRS can be enhanced when the pump frequency is near a resonance and the virtual state is replaced by a real state, which is resonance Raman scattering (RR).

SRS requires time-overlap of laser light at both the pump and signal frequencies. This is possible using separate laser pulses or a single laser pulse with broad bandwidth. The medium can be excited or de-excited, as SRS exchanges energy between the medium and the pump and signal photons. For example, the bandwidth of ultrashort pulses can cover

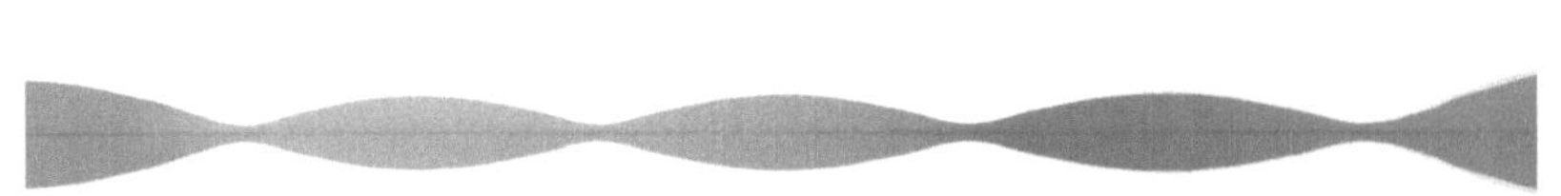

Figure 2.5: A cartoon of filamentation, spectral broadening, and self-seeding.

many Raman transitions involving different closely spaced rotational states of a molecule, so SRS can create a coherent superposition of different rotational states. The resulting rotational wave packet modulates molecular alignment, as discussed in Ch. 5.

If the lower states in Figure 2.4 are far apart, SRS can provide gain at signal frequencies relatively far from the pump frequency; however, gain is accompanied by absorption and is only available when the pump and signal fields overlap in time. In contrast, gain from an inverted system (Section 2.3.1) does not require time overlap or simultaneous absorption.

2.4 Air lasing in N_2^+

The air laser describes an idealized tool for remote sensing that would provide laser pulses remotely starting from air [15, 16]. First, the air laser requires a well-understood gain medium. This scientific challenge has dominated air lasing research efforts so far. Once the gain medium is operational and fully characterized, optimizing and manipulating it become technical challenges.

The technical challenges are significant due to filamentation in air, which may be the delivery mechanism of the air laser. A simple view of filamentation is the balance between self-focusing and plasma defocusing [17, 18, 19, 20]. If the laser is above the critical power in air, it can self-focus. Self-focusing increases the intensity by reducing the size of the laser, so it begins to ionize the medium and generate plasma. The plasma defocuses the beam, which may be sufficiently powerful to self-focus again. The interplay of these effects can allow a focused beam to propagate over distances much longer than the Rayleigh length [21], but it is sensitive to the laser, focusing, and environmental conditions [22, 23, 24, 25, 26, 27, 28].

Figure 2.5 illustrates filamentation as a balance between self focusing and plasma defocusing, which may generate a sausage-like shape. This interplay limits the energy in a single filament to a clamping intensity of about 5×10^{13} W/cm^2 [19, 26, 29], and a powerful beam can split into multiple filaments. The focused laser accumulates a lot of nonlinear phase during propagation over macroscopic distances in air filaments, which can significantly broaden the spectrum of the pulse [30, 31, 32, 33]. Filamentation influences the laser duration, spectrum, spatial profile, and polarization [34, 35, 36, 37, 38, 39, 40, 41, 42, 43].

Air lasing research has explored a number of molecules and atoms that provide gain during filamentation. For example, neutral molecular nitrogen provides gain that is collisionally-pumped like the traditional nitrogen laser [44, 45, 46, 47, 48, 49, 50, 51, 52, 53]. In filamentation, electrons in the plasma collide with and excite nitrogen. Unfortunately, the excited nitrogen molecules quickly lose their energy by dissociating molecular oxygen, which removes gain at concentrations comparable to air. As a result, neutral nitrogen is a poor gain medium for the air laser.

Similarly, atomic nitrogen and oxygen produce gain in air using an ultraviolet pump pulse [54, 55]. These are also poor candidates because the pump pulse cannot propagate through air beyond a couple hundred meters due to absorption. Carbon dioxide also provides gain, but it is not a large fraction of air [56]. Fortunately, N_2^+ provides gain in the visible and soft UV using a large range of near-infrared pump pulse wavelengths during air filamentation.

The gain in N_2^+ was first reported after focusing a 1.9 μm femtosecond laser in air [2]. A bright and narrow peak appeared on the spectrum of the fifth harmonic, and the peak exhibited the same linear polarization as the input laser. The location of the peak corresponded to the transition between the ground vibrational states of the $B^2\Sigma_u^+$ and $X^2\Sigma_g^+$ electronic states of N_2^+ shown in Fig. 2.6, and the intensity of the peak scaled exponentially with the plasma length. The first publication also demonstrated these characteristics of lasing for four other transitions involving different vibrational states by tuning the harmonic seeding the emission to overlap each transition. In most cases, the center wavelength of the harmonic self-seed was adjusted using the pump wavelength, which also demonstrated a broad pump wavelength range from 1.4 to 2.0 μm.

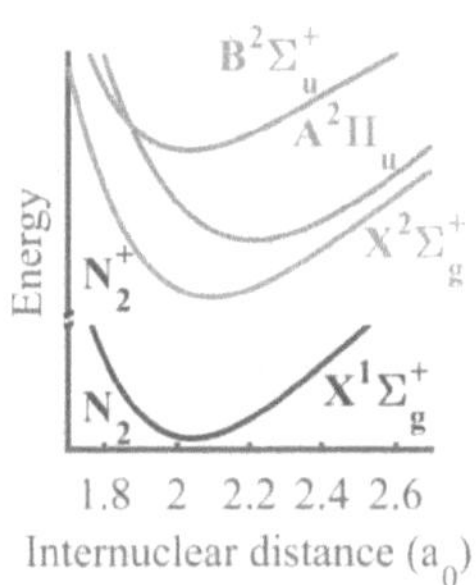

Figure 2.6: Potential energy as a function of internuclear distance for the ground state of the neutral nitrogen molecule and the first three states of the cation.

Nine years after it was first reported, there is no consensus about the origin of gain [16, 57, 58]. Consensus may be hindered by an abundance of measurements using different conditions with differing results. The gain is available for a variety of pump laser wavelengths [2,59,60,61,62,63,64,65,66], lasing wavelengths [2, 4, 59], gas pressures [59,67,68,69], and gas mixtures [2, 4, 59, 69]. In contrast with the first report, many follow-up experiments including ours use an 800 nm pump pulse due to the relative abundance of titanium-sapphire based laser systems and the strong air lasing signal at this pump wavelength.

Some measurements of gain are self-seeded due to the continuum generated by the pump pulse [2, 59, 60, 62, 70, 71, 72, 73, 74, 75, 76]. Usually, an external probe pulse measures gain separately from the pump pulse [2, 67, 77, 78], which sometimes occurs after the continuum from the pump pulse already self-seeds gain [69, 79]. The probe pulse often measures gain that falls as a function of delay with an exponential-like decay over picoseconds after the pump pulse [4, 65, 69, 79, 80, 81, 82, 83, 84, 85, 86, 87, 88].

Rapid oscillations are usually superimposed on the decay due to rotational wave packets in the $B^2\Sigma_u^+$ and $X^2\Sigma_g^+$ states that modulate the alignment of each state [, , , ,] and, equivalently, modulate the transitions between the states [, , , , ,]. Propagation of the probe pulse through the dynamic gain medium further increases the complexity of the modulations [,].

The exponential-like decay of gain over tens of picoseconds is consistent with the decay of population inversion due to collisions [,]. These collisions may be the source of rotational decoherence [ ]. The fluorescence lifetime of the $B^2\Sigma_u^+$ state is much longer. It is sub-nanosecond for ultrashort pulses, which is determined by collisional depopulation [ ] and includes an additional faster component [, , ,]. The lifetime in low-pressure RF plasma is even longer. It is on the order of tens of nanoseconds [ ].

In some conditions, the exponential-like decay is replaced by a brief spike of gain when the pump and probe pulses overlap [, , ,]. In other conditions, gain is replaced by absorption using a different wavelength of the pump [,] or probe [ ], a polarization-modulated pulse [ ], or an additional pulse [ ].

Experiments that time-resolve the probe pulse at a fixed delay [ ] or the continuum self-seed [ ] measure emission that also lasts for many picoseconds after the probe pulse. In some conditions, several characteristics of the emission are consistent with a collective many-body effect known as superradiance [, ,]. In most conditions, the emission exhibits complicated structure in time due to modulated gain during propagation and beating between transitions involving different rotational states [, , , , ,].

Ionization will produce some population in the $B^2\Sigma_u^+$ state initially [,]. The middle $A^2\Pi_u$ state provides a destination for the population in the $X^2\Sigma_g^+$ state using pump pulses near 800 nm, which could contribute to population inversion [, , , , , , , ,]. The slight dependence of gain on the pump ellipticity [,] is consistent with the transfer of a small amount of population to the $B^2\Sigma_u^+$ state via inelastic scattering of an electron during recollision [, , ,], as discussed in Ch. 4. Alternatively, it may be explained by post-ionization coupling between the $X^2\Sigma_g^+$ and $A^2\Pi_u$ states [,], which we utilize in Ch. 6. Using an additional pulse, rapid modulations of gain and emission as a function of delay demonstrate electronic and vibrational coherence [, , , ,].

A population inversion between the ground and excited states is one explanation for gain [, , ,], but a measurement of gain does not imply population inversion. Alternatively, transient [,] or partial [, ,] rotational inversion could produce gain in some conditions. The former is similar to population trapping that modulates the fluorescence of both neutral and ionic nitrogen [ ]. The latter may be highly sensitive to the chirp of the excitation pulse [ ]. Wave mixing may also produce gain in other conditions [, , , ,]. The air laser involves contributions from all of these effects, which is partially responsible for the sprawling research directions in the literature.

Understanding gain is one problem, but control of gain and emission are also necessary to create an air laser device. Fortunately, the volume of research targeting coherent and other control has increased with our understanding of air lasing over time. An early example used pulse shaping techniques to form a pulse train at 1030 nm with two main pulses separated by a controlled time delay, and they observed enhanced gain during filamentation for delays corresponding to partial rotational revivals in the ion [116]. More recently, tunable elliptically polarized lasing in a pump-probe geometry was demonstrated and attributed to strong birefringence in the gain medium [117]. Furthermore, an alignment pre-pulse can enhance gain [118]. Polarization-modulated pulses can also enhance gain using multistate coupling that is otherwise inaccessible due to molecular alignment [98, 110, 119]. It is possible to halt lasing emission using a control pulse in order to tune the emission duration [5, 120], and a control pulse can annihilate rotational wave packets in the states of the ion [5]. One of the most essential components of an air laser device, backwards lasing, is also under investigation [121].

Chapter 3

Gain in the Jet

Chapter 2 introduced a number of linear and nonlinear effects that distort a pulse in space and time during propagation. Figure 3.1 shows the result of focusing a powerful 800 nm laser pulse in extremely uncontrolled conditions. While the image is blurry and taken from an angle, it is clear that the pulse has broadened to the blue end of the visible spectrum from the infrared. In addition, the spatial profile is nonuniform and messy. My colleague calls it "the flamethrower" in part because the small fluctuations of the environment or the laser make the messy spatial profile move around like flames. Figure 3.1 shows extreme conditions, but gain exists in these conditions. It is tempting to study gain during filamentation in air. In that case, while the experiments may be more controlled than Fig. 3.1, the laser pulse still changes as it propagates and creates the gain medium nonuniformly.

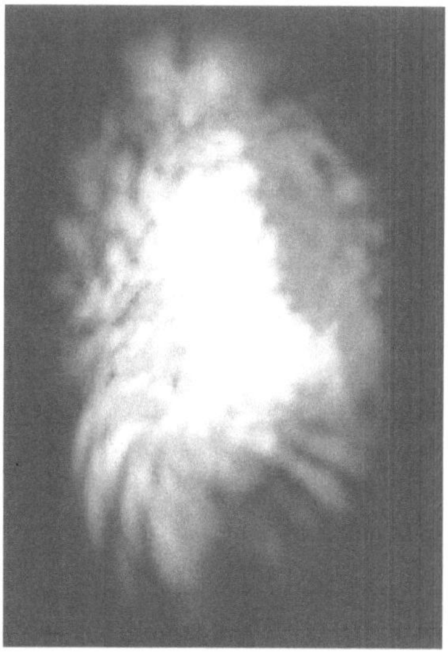

Figure 3.1: Outside of the jet: a photo of emission from plasma in air on a black screen from the side. The center is saturated.

Our approach minimizes propagation distance using a narrow gas jet in vacuum so that the laser pulses are relatively unchanged by the interaction. This method isolates gain from filamentation. It also prevents self-seeding, so a separate probe pulse measures gain and resolves dynamics as a function of delay.

Figure 3.2 shows a diagram of this experiment. A dichroic mirror (DM) in a Mach-Zehnder interferometer recombines the 800 nm pump pulse with the 400 nm second harmonic probe pulse colinearly with a controlled delay between them (Probe delay). The pump pulse determines zero probe delay, and positive delays correspond to the pump pulse arriving first. Recombining the pulses within the chamber increases the sensitivity of the measurement by removing weak spectral modulations due to air outside of the vacuum chamber in some conditions.

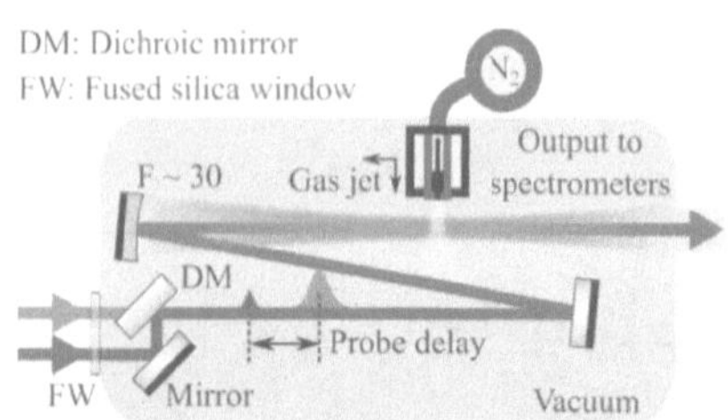

Figure 3.2: A diagram of the experiment to measure gain in the gas jet.

Two methods determine zero delay. First, inserting a frequency-doubling crystal in the pump beam produces interference between the probe pulse and the second harmonic of the pump pulse when they are overlapped in space and closely separated in time [7]. We monitor fringes in the spectrum due to this spectral interference. If the two second harmonic pulses are identical and perfectly overlapped, the spectrum is unmodulated, while the fringes move through the spectrum at small delays. The second method requires measuring high harmonics generated by the pump pulse. When the second harmonic probe pulse overlaps the pump pulse, high harmonics are detected at even integer multiples of the pump wavelength as discussed in Sec. 4.1 [122].

The pulses are focused using a spherical focusing mirror with a focal length of 30 cm. The diameter of the laser beams at the focusing mirror is approximately 1 cm. We ensure spatial overlap in the focus by iteratively aligning the probe pulse to the pump pulse prior to the focusing mirror and then in the far field. To aid in visually overlapping the beams, temporary apertures before the vacuum chamber produce Airy patterns with clear rings that identify the centers of the beams.

The gas jet is on linear stages to move it in three dimensions. The nozzle position is as close to the focus as possible. This position is determined by locating the maximum ionization without clipping the laser beam spatially. Maximum ionization is identified visually using the intensity of the fluorescence from the plasma and also on an oscilloscope connected to a large copper mesh near the focus that measures ions created by the strong laser field.

An Even-Lavie pulsed valve creates the gas jet [123]. The nozzle is cylindrical with a diameter of 200 μm. The valve housing extends approximately 1.5 cm around the center of the nozzle, which limits the distance between the nozzle and the focus to at least a few hundred micron. Nitrogen gas is supplied to the valve with a backing pressure of 5 to 7 bar.

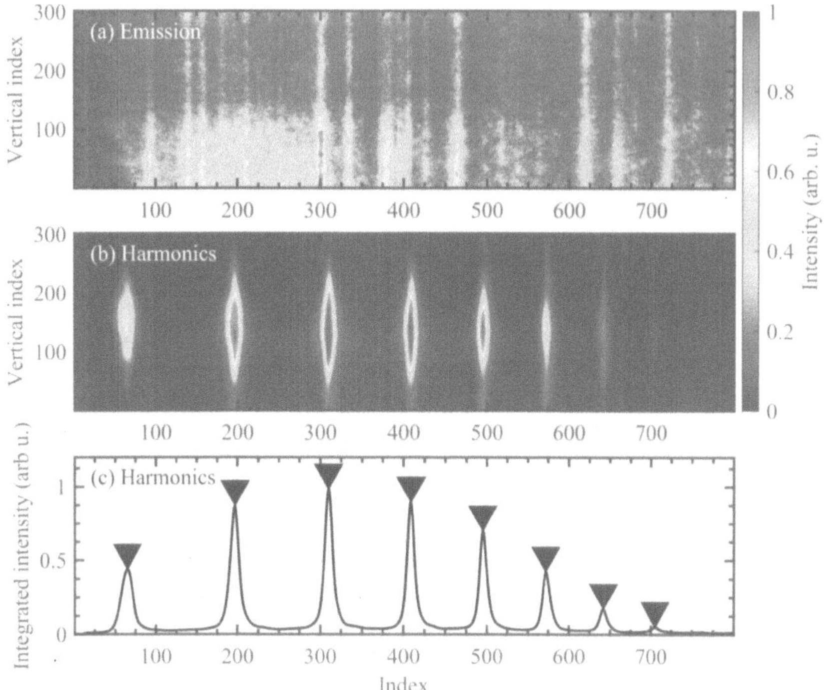

Figure 3.3: N_2 (a) Image of the MCP showing emission lines where index refers to the pixel number, (b) image of the MCP showing high harmonics, (c) Integrated intensity of high harmonics and triangles indicating the peak locations.

3.1 Cluster formation

Clusters of molecules can form in supersonic gas jets. The Hagena parameter quantifies the prominence of clusters [ ]. For sonic nozzles like the cylindrical shape, the Hagena parameter

$$\Gamma^* = k \frac{d^{0.85} P_0}{T_0^{2.29}} \tag{3.1}$$

depends on the diameter of the jet d in micron, gas temperature T_0 in Kelvin, and the backing pressure P_0 in mbar. $k = 528$ is the condensation parameter for nitrogen.

The average cluster size $\langle N \rangle$ for sonic nozzles is

$$\langle N \rangle = 33 \left(\frac{\Gamma^*}{1000} \right)^{2.35} . \tag{3.2}$$

when $\Gamma^* \approx 7300$. Cluster formation is not significant below $\Gamma^* = 1000$, where the average cluster size is $\langle N \rangle \approx 33$ [125]. In our conditions, Γ^* is 550 to 750, so cluster formation is negligible. For comparison, the average distance between nitrogen molecules in our conditions is on the order of tens of nanometers, while the typical distance in a cluster is on the order of several Angstroms [124, 126].

3.2 Intensity Calibration

After the focus, a spectrometer can measure part of the extreme ultraviolet (XUV) electromagnetic spectrum. The gas jet and spectrometer are part of a high harmonic beamline. The spectrometer measures the intensity of the pump pulse because the peak intensity of a laser pulse determines the highest harmonic in the spectrum according to the high harmonic cut-off law [1, 127]. The spectrometer consists of an adjustable slit, a physical XUV grating, and a microchannel plate (MCP) detector within the vacuum chamber. A camera and imaging system capture images of the MCP from outside of the chamber. The groove spacing of the grating and its position relative to the slit and detector determine the spectrometer range and resolution.

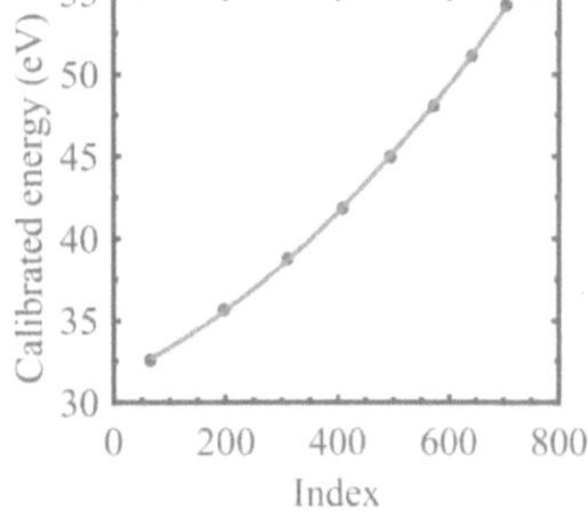

Figure 3.4: Calibrated axis assuming 21 is the lowest harmonic order visible in Fig. 3.3.

Figure 3.3(a) and (b) show images of the MCP using an intense circularly and linearly polarized pump pulse in nitrogen, respectively. The images are rotated and cropped, which corrects the angle of the camera and removes dark regions on and around the MCP. Elliptical or circular polarization suppresses high harmonic generation in atoms and simple molecules like N_2 [128, 129], as discussed in Ch 4. In that case, it is possible to detect the emission lines in Fig. 3.3(a). We use the harmonics and emission lines, along with the expected positions of emission lines from literature [130], to determine the photon energy axis calibration of the spectrometer.

Figure 3.3(c) shows the intensity of the high harmonics in Figure 3.3(b) integrated over the vertical index. Triangles indicate the positions of the peaks in the high harmonic spectrum, which roughly constrains the photon energy axis because the peaks are odd harmonics of the pump pulse. Even harmonics are absent due to the symmetry of the pump laser cycle [122]. Even harmonics can also be generated when the second harmonic probe pulse is present because it breaks the symmetry of the laser cycle.

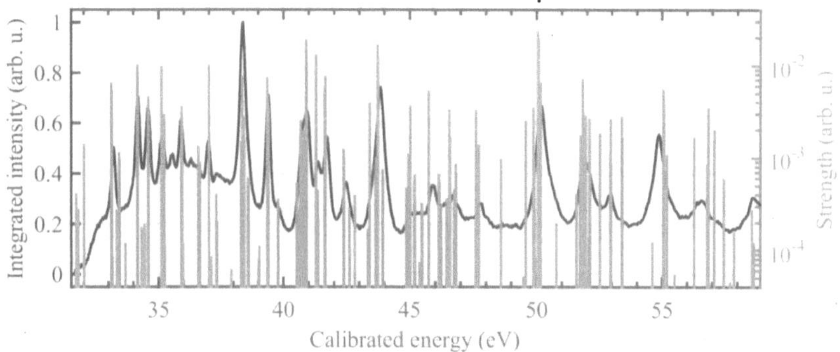

Figure 3.5: Integrated intensity of N_2 emission in Fig. 3.3(a) using the guessed axis in Fig. 3.4 (left-hand axis), and the known transition locations (right-hand axis).

The central photon energy of the pump pulse and the lowest harmonic determine the photon energy axis. For example, Fig. 3.4 shows the calibration assuming that the lowest visible harmonic order is 21 and the photon energy of the driving laser field is 1.55 eV. Lower orders miss the MCP. A superimposed polynomial fit can extrapolate the data.

If the lowest visible harmonic is unknown, the correct value can be determined by comparing the emission lines with the expected positions using the guessed axis. When the guesses are close, the comparison is good visually like Fig. 3.5. This calibration is imperfect but sufficient to determine the energy of the highest harmonic generated. For additional precision, the polynomial fit to the harmonic positions in Fig. 3.4 can be replaced by a nonlinear fit of the grating equation

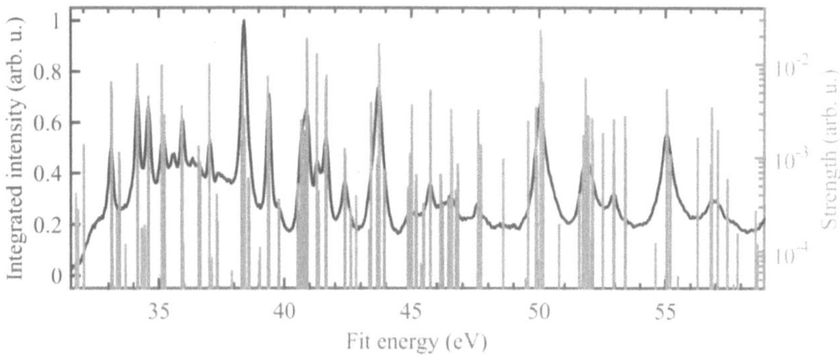

Figure 3.6: Like Fig. 3.5, but with the energy axis extracted from a nonlinear fit using the grating equation given by Eq. 3.3.

$$sin(\theta_i) + sin(\theta_o) = \frac{m\lambda}{d} \tag{3.3}$$

to the known energies and corresponding indices in Fig. 3.5, where θ_i is the incident angle, θ_o is the diffracted angle, m is the diffraction order, d is the groove spacing of the grating, and λ is the diffracted wavelength. Fig. 3.6 shows this result.

This calibration shows that the highest order harmonic in Figure 3.3(b) is 35. According to the high harmonic cut-off law, the highest photon energy E_{max} is

$$E_{max} = I_P + 3.17U_P, \tag{3.4}$$

where $I_P = 15.6$ eV is the ionization potential of nitrogen. The ponderomotive energy

$$U_P = \frac{2e^2}{c\epsilon_0 m_e}\frac{I}{4\omega_0^2} \tag{3.5}$$

is related to the intensity I of the pump pulse by physical constants and the frequency of the laser. For 800 nm in nitrogen, $E_{max} = 54.3$ eV for the 35th order corresponds to a peak intensity of $I = 2 \times 10^{14}$ W/cm^2.

A half wave plate and polarizer control the pump pulse energy before the vacuum chamber. The intensity in the focus should scale linearly with pulse energy. To improve the intensity calibration, we determine the intensity according to the high harmonic cut-off law at a series of pulse energies and perform a linear regression on the data.

The duration of the pump pulse is approximately 35 fs outside of the vacuum chamber. We optimize the position of the grating in the compressor of the chirped pulse amplification (CPA) laser system to minimize the pump pulse duration in the focus by maximizing strong-field ionization of N_2. As a result of this and the frequency-doubling process, the probe pulse is chirped and longer than the minimized pump pulse duration. We estimate the probe pulse duration to be 80 fs. This estimation is consistent with the range of probe delays where even high harmonics are detected (*e.g.* Fig. 5.9(b)).

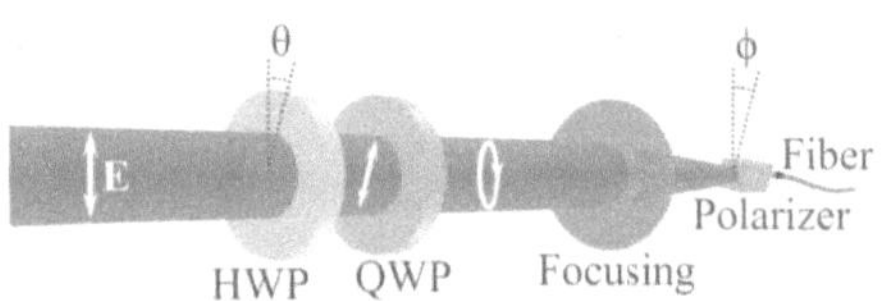

Figure 3.7: Diagram of the components to control and measure the pump polarization state.

3.3 Ellipticity calibration

The ellipticity of the pump polarization controls the efficiency of high harmonic generation in Figure 3.3(a) and (b). This type of measurement will return in Ch. 4, where the control

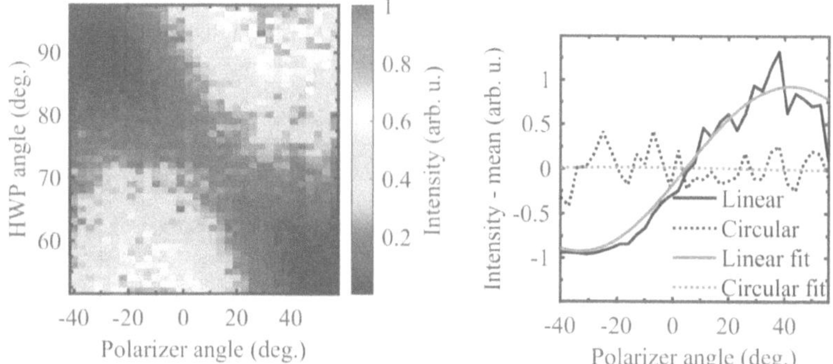

Figure 3.8: Intensity transmitted through the polarizer as a function of half wave plate (HWP) and polarizer angle. Zero polarizer angle and zero half wave plate angle correspond to raw stage positions.

Figure 3.9: Lineouts from Fig. 3.8 that are near linear and circular polarization.

of high harmonic signal verifies that gain in N_2^+ exists without inelastic scattering during electron recollision. Half and quarter wave plates control the ellipticity and orientation of the laser polarizations, which are measured near the focus.

To create and calibrate the ellipticity of the pump pulse in the gas jet, a half wave plate (HWP) on a rotation stage rotates the input vertical polarization of the pump pulse before it reaches a quarter wave plate (QWP), as shown in Fig. 3.7. The fixed quarter wave plate generates elliptical polarization with the major axis horizontal or vertical. The beam propagates to the focus, where a polarizer on a rotation stage attenuates the beam to calibrate the polarization state. Figure 3.8 shows the integrated intensity transmitted through the polarizer measured on a spectrometer as a function of the angle of the half wave plate and the polarizer. For a linearly polarized pulse, rotating the polarizer produces a sinusoidal oscillation between the peak intensity and zero intensity.

If the input polarization is vertical and only the half wave plate, quarter wave plate, and polarizer are in the beam, the Jones vector after these elements

$$\begin{pmatrix} E_x(t) \\ E_y(t) \end{pmatrix} = \frac{1}{2} \begin{pmatrix} \sin 2\theta\,(1+\cos 2\phi) - i\cos 2\theta \sin 2\phi \\ \sin 2\theta \sin 2\phi - i \cos 2\theta\,(1-\cos 2\phi) \end{pmatrix} \tag{3.6}$$

only depends on the angle of the polarizer ϕ and the angle of the half wave plate θ shown in Fig. 3.7. The half wave plate angle determines the ellipticity

$$\epsilon(\theta) = \begin{cases} |\tan 2\theta| & -\frac{\pi}{8} \leq \theta \leq \frac{\pi}{8} \\ |\cot 2\theta| & \frac{\pi}{8} \leq \theta \leq \frac{3\pi}{8} \end{cases} \tag{3.7}$$

after the quarter wave plate, where $\epsilon = 0$ and 1 correspond to $\theta = 0°$ and $\pm 22.5°$, respectively. Ignoring the polarization sensitivity of the spectrometer and other optics, the intensity is proportional to the magnitude of the Jones vector

$$|E(t)| = \frac{1}{2}\left(1 - \cos 2\phi \cos 4\theta\right), \tag{3.8}$$

so the ellipticity $\epsilon(\theta)$ determines the modulation depth as a function of the polarizer angle ϕ. Using circular polarization, the transmission through the polarizer should be constant as a function of polarizer angle. The periodicity of Fig. 3.8 is similar to the periodicity in Eq. 3.8, and it continues outside of the region on the figure.

We fit a sinusoidal function to the intensity modulation at each half wave plate angle to extract the modulation depth and determine the ellipticity. Figure 3.9 shows horizontal lineouts from Fig. 3.8 at half wave plate angles near circular polarization (72 deg) and horizontal linear polarization (90 deg). It also shows the corresponding fits. The method does not enforce the expected periodicity of the half wave plate, nor does it constrain the angle of the major axis of the measured polarization state, so it can account for modified ellipticity or a rotated ellipse. By measuring the polarization near the focus, this method accounts for intermediate optics in the path.

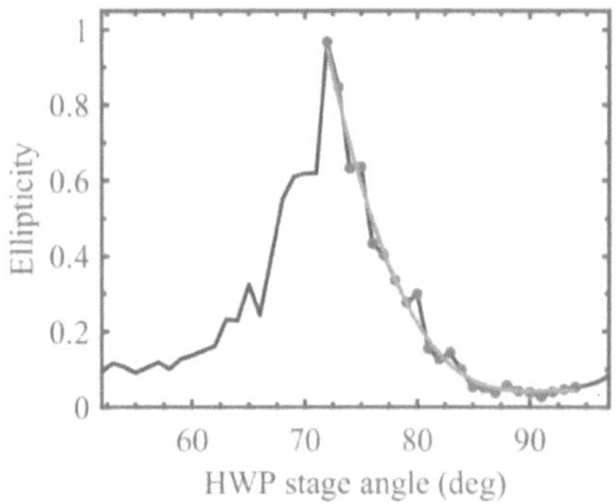

Figure 3.10: Ellipticity as a function of half wave plate angle extracted from fits like those in Fig. 3.9 to the data in Fig. 3.8.

Figure 3.10 shows the ellipticity determined from the fits to the data in Fig. 3.8. A polynomial least-squares fit is superimposed, which extrapolates and smooths the calibrated ellipticity axis.

The angle of the quarter wave plate can also control the ellipticity. In that case, the major axis of the polarization ellipse also rotates with the quarter wave plate. This is particularly undesirable in pump-probe experiments using a linearly polarized probe pulse because the angle between the polarizations is important in many processes, but the combination of a static quarter wave plate following a rotating half wave plate can preserve the orientation of the pump pulse relative to the probe pulse.

3.4 Detection and analysis

After the focus in Fig. 3.2, a translation stage can remotely insert a mirror to redirect the pulses away from the XUV spectrometer and out of the vacuum chamber. The pulses are filtered and attenuated, and refocused into a fiber spectrometer. Figure 3.11 shows the

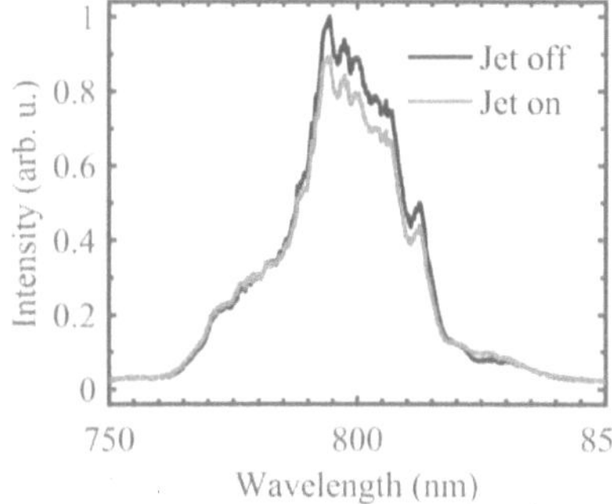

Figure 3.11: The spectrum of the pump pulse with the gas jet on and off.

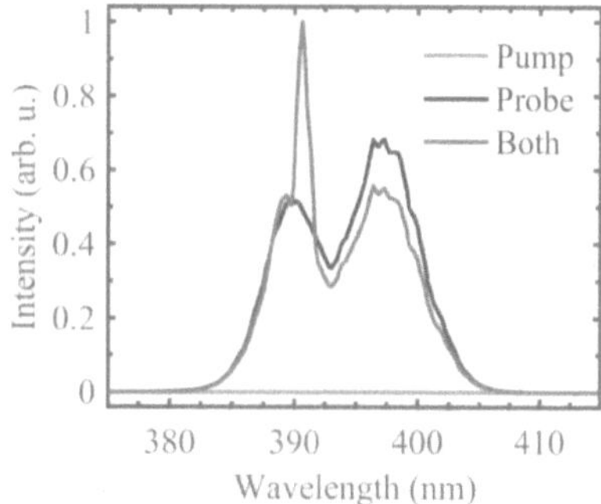

Figure 3.12: The spectrum with the pump pulse, the probe pulse, or both together.

spectrum of the pump pulse with the gas jet off and on. The pump pulse is blue-shifted with the jet on due to ionization [81, 82], but self-phase modulation is insignificant.

Spectral filters and a dichroic mirror remove the pump pulse and isolate the probe pulse after the focus. An Ocean Optics Maya Pro 2000 fiber spectrometer measures the spectrum of the probe pulse. Figure 3.12 shows the spectrum of the probe pulse with and without the pump pulse, and the spectrum with only the pump pulse for reference. Occasionally, an integration sphere reduced the sensitivity of alignment into the spectrometer to reproduce important measurements.

There is no intensity around 400 nm with only the pump pulse in Fig. 3.12. This also confirms that spectral broadening of the pump pulse is insignificant due to the limited propagation distance. The pump pulse generates a small amount of second harmonic on optics like wave plates that is removed using filters and a dichroic mirror. The pump pulse can also generate second harmonic during the interaction in the gas jet due to molecular alignment, but only a small amount appears using uncommonly high intensities. In addition, the bandwidth of the second harmonic that is generated from molecular alignment is narrow, and it does not overlap the transitions that provide gain. This prevents self-seeding, so the probe pulse measurement is isolated.

Figure 3.12 shows the spectrum of the probe pulse alone. The angle of the 200 μm-thick frequency-doubling BBO crystal influences the spectrum of the probe pulse by tuning the phase matching condition. A half wave plate and polarizer set the intensity on the BBO crystal, and self-phase modulation can introduce additional frequencies at high input intensity. This enables the measurement of gain at transitions farther from the center of the second harmonic spectrum. After the BBO, a polarizer converts the probe polarization to vertical. With both the pump and probe pulses, gain appears around 391 nm [$\mathrm{B}^2\Sigma_\mathrm{u}^+$ ($\nu = 0$) $\rightarrow$ $\mathrm{X}^2\Sigma_\mathrm{g}^+$ ($\nu = 0$)] in Fig. 3.12. This spectrum represents a measurement of gain at a single probe delay.

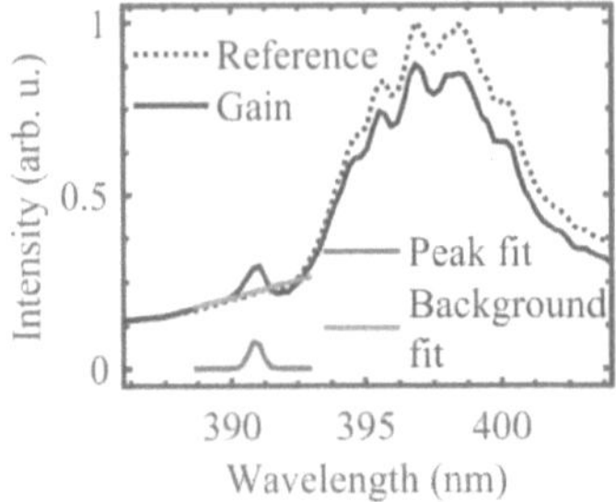

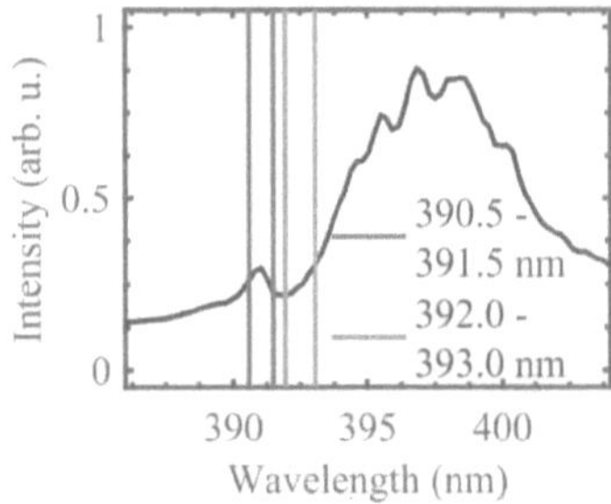

Figure 3.13: The probe pulse with and without the pump pulse, and fits of the peak and the probe (background).

Figure 3.14: Like Fig. 3.13, but highlighting integrated regions using vertical lines.

The ratio of the output intensity to the input intensity is related to the average gain. Figure 3.13 shows gain in different experimental conditions from Fig. 3.12. A dotted line indicates the input probe spectrum. Effects other than gain influence the probe pulse, such as absorption and plasma defocusing. Thermal or mechanical drift of the optics can also change the spectrum slowly over the duration of a measurement. Therefore, it is necessary to monitor the input intensity as a function of probe delay. A polynomial fit of the probe and a Gaussian fit of the peak are superimposed on Fig. 3.13, which can separate the variation of gain from the variation of the probe intensity. We benefitted from fitting the data this way, as it lessened the effects of noise and drift in some conditions. For formal data, we prefer an approach that requires less intervention.

Instead of fitting the input and peak at each probe delay, it is faster and simpler to integrate over two regions. The input intensity is unknown without the background fit, but it is approximately the integrated intensity in the same region at negative probe delays. This is imperfect because it does not capture the variation of the input intensity. Monitoring a nearby non-amplified region of the spectrum can account for these changes.

Figure 3.14 highlights the amplified region of the spectrum and a nearby non-amplified region. Figure 3.15 shows the integrated intensity in these two regions as a function of probe delay. Both regions vary as a function of delay, and they are normalized to their average values at negative probe delays. For the amplified region (390.5 – 391.5 nm), this represents the ratio of the output intensity I_{out} at positive delays to the input intensity I_{in} given by the average intensity at negative delays. The natural logarithm of this quantity is the raw gain-length product

$$\mathrm{gL}_{raw} = \log \frac{I_{out}}{I_{in}}, \tag{3.9}$$

which is still sensitive to the variation of the input intensity.

The non-amplified region (392.0 – 393.0 nm) represents false gain and absorption. It is the ratio of the actual input intensity $I_{in,act}$ at positive delays to the expected input

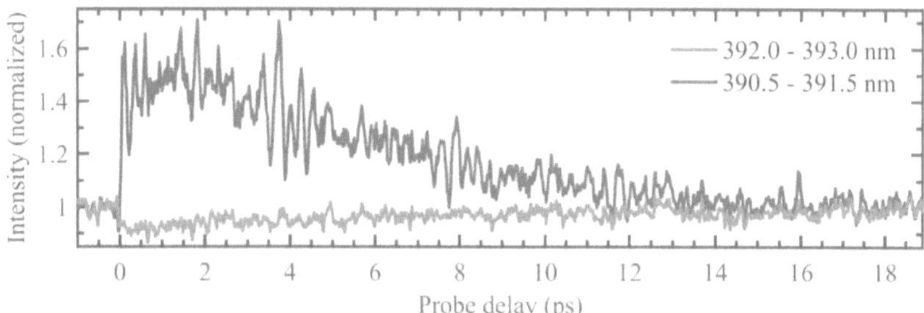

Figure 3.15: The integrated intensity in the regions of Fig. 3.14 as a function of probe delay.

intensity $I_{in,exp}$ given by the average value at negative delays. The non-amplified region should be unity if the input spectrum is unmodified. We correct the amplification ratio by dividing the raw amplification ratio by the false gain amplification ratio. The resulting gain

$$\text{gL} = \log\left(\frac{I_{out}}{I_{in}}\frac{I_{in,exp}}{I_{in,act}}\right) \tag{3.10}$$

is less sensitive to changes of the probe spectrum.

Figure 3.16 shows the gain length product as a function of probe delay. Unless otherwise noted, gL refers to gain at 391 nm. Gain appears at positive delays and decays over tens of picoseconds, and the decay is modulated. These features are explored in detail throughout this book. Shot-to-shot variation and the slow drift of the probe laser intensity occur due to the variation of the input laser and the heating of optics. By scaling the reference spectrum according to the instantaneous intensity of an unmodified spectral region, the

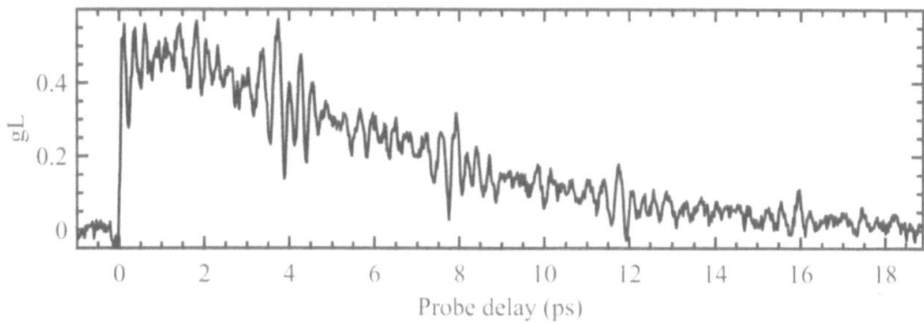

Figure 3.16: The natural logarithm of the ratio of the integrated intensity in the amplified region to the integrated intensity in the non-amplified region from Fig. 3.15.

correction factor in Eq. 3.10 accounts for these variations.

3.5 Probe intensity

We estimate the probe intensity by scaling the calibrated intensity of the pump pulse using the relative energy of the two pulses while accounting for the different wavelength, beam size, and pulse duration. As a result, we estimate that the peak intensity of the probe pulse is less than 10^{10} W/cm^2. The intensity of the probe pulse is important because we prefer to measure the small-signal unsaturated gain and also minimize cascaded resonant Raman rotational excitation. Small-signal gain can be directly verified by scaling the probe intensity. Figure 3.17 shows gain measured using two probe pulse intensities that differ by a factor of two. The spectra for both probe intensities are normalized to the average input intensity in the gain region. Both probe pulses measure the same gain at 391 nm despite differing in intensity by a factor of two, which is consistent with a measurement of the small-signal gain.

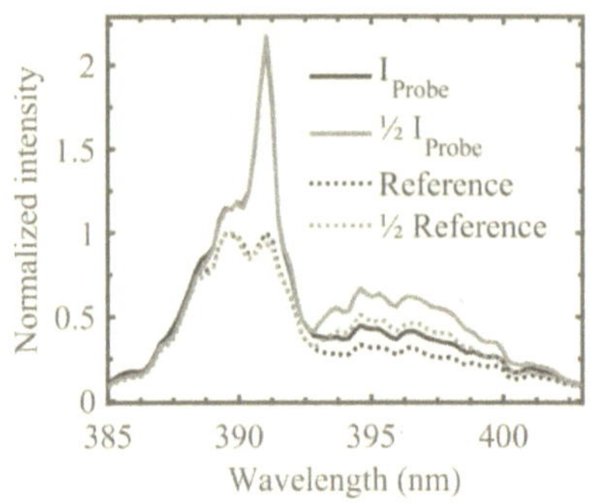

Figure 3.17: Gain with probe intensities that differ by a factor of two and references (dashed lines) excluding the pump. The intensity is normalized to the average reference intensity from 389.5 to 392.6 nm

3.6 Summary

We study gain using a narrow nitrogen gas jet in vacuum to minimize propagation distance and nonlinear effects. The pump pulse is relatively unchanged as a result. The probe pulse makes an isolated measurement of gain at 391 nm. An XUV spectrometer allows us to measure high harmonics to calibrate the pump intensity or determine time overlap with the probe. We calibrate the ellipticity of the pump near the focus. We measure the small-signal gain. Integrating over the amplified region of the spectrum provides gain as a function of probe delay.

Chapter 4

The Role of Recollision

4.1 Recollision

Recollision is the third step of the semi-classical three step model of strong field processes like high harmonic generation [,]. The first step is tunnel ionization when the strong laser modifies the potential of the atom or molecule and an electron tunnels through the classically forbidden barrier. During the second step, the laser accelerates the electron classically from rest. During the third and final step, the electron returns to the parent ion.

When the electron returns to the parent ion, it can recombine and release an extreme ultraviolet (XUV) photon. In high harmonic generation, this produces bursts of XUV radiation each half-cycle of the driving optical field. Under the right conditions, this XUV spectra forms an attosecond pulse train. High harmonics appeared in Ch. 3 as a tool to calibrate the intensity of the pump pulse. Figure 3.3(b) shows only odd harmonics of the pump pulse because the XUV pulses in the train are separated by half of an optical cycle and even-order harmonics destructively interfere. Even-order harmonics can be generated by introducing a second-harmonic field, which breaks the half-cycle symmetry of the driving field so every other pulse in the train is slightly different [ ]. Consequently, Ch. 3 discussed the generation of even-order harmonics as a tool to determine the time overlap of the pump pulse and the second harmonic probe pulse.

Additional processes occur when the electron wave packet recollides with the parent ion in the third step. For example, recollision contributes to double ionization in N_2 [ ]. The electron can also scatter inelastically during recollision and excite the parent ion. Chapter 3 introduced an isolated measurement of the gain in N_2^+ between the $B^2\Sigma_u^+$ and $X^2\Sigma_g^+$ states. This gain suggests that the $B^2\Sigma_u^+$ state is populated. It has been suggested that electron recollision may contribute to the excited population [, , , ,].

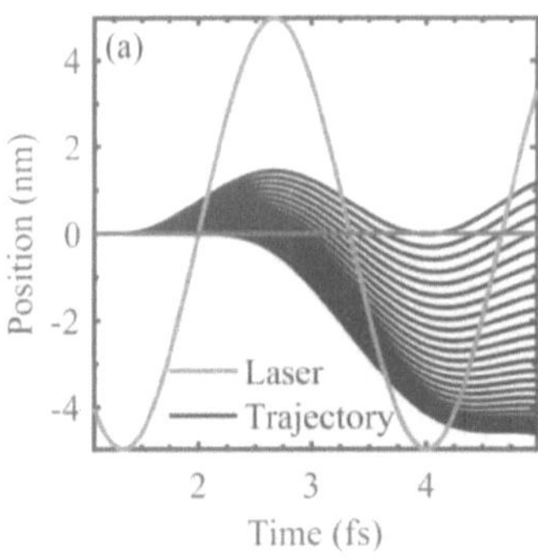

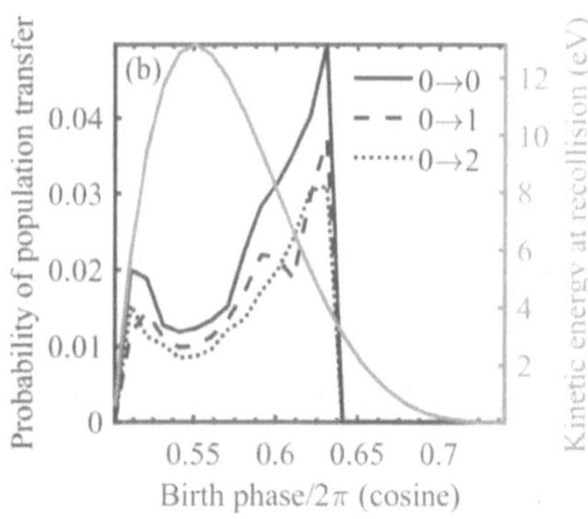

Figure 4.1: (a) The laser field and classical electron trajectories [ ]. (b) The probability of population transfer from the $X^2\Sigma_g^+(\nu_i = 0)$ state to the $B^2\Sigma_u^+(\nu_f = 0, 1, 2)$ states in N_2^+. The red line (right hand axis) shows the kinetic energy of the electron at the time of recollision. The laser is 800 nm and 7×10^{13} W/cm^2

4.2 Semiclassical model

The energy-dependent cross section for electron excitation to the $B^2\Sigma_u^+$ state is known [ ], so attoscience provides a variety of tools to calculate the probability of excitation during recollision. Although not a replacement for a rigorous theoretical model, a basic semiclassical model is a simple means of estimating the population transfer and developing intuition. In the semiclassical calculation, the classical motion of electrons as point particles under the influence of the laser field provides the distribution of recolliding electron energies. Figure 4.1(a) shows the laser field and recolliding electron trajectories, and the red line on (b) shows the recolliding energies. The electron enters the continuum at time t_b, known as the time of birth, and returns to the parent ion at time t_r with kinetic energy $\epsilon(t_b)$.

The initial electron wave packet is highly localized and spreads out in space over time. The model approximates this quantum diffusion by assuming an exponentially decaying wave function

$$\psi(r,t) = \exp\left\{-\frac{r}{r_\perp(t)}\right\} \tag{4.1}$$

and a transverse radius of the electron wave packet after tunneling of $r_\perp(t_b) = 0.32$ nm, which is comparable to the dimensions of the nitrogen molecule. The transverse velocity of the expansion $v_\perp = 2.3$ nm/fs can be estimated using the uncertainty principle [ ]. The final radius $r_\perp(t_r) = r_\perp(t_b) + v_\perp\,[t_r - t_b]$ determines the overlap with the ion. The probability of recollision

$$P(t_b) = \iint_\sigma |\psi(r,t_r)|^2 \mathrm{d}A \tag{4.2}$$

is the integral of the normalized wave function upon return within a circle with area equal to the cross section centered on the ion, which also depends on the electron energy $[\sigma = \sigma(\epsilon) = \sigma(t_b)]$.

Figure 4.1(b) shows the probability of inelastic scattering between the $X^2\Sigma_g^+(\nu_i = 0)$ and $B^2\Sigma_u^+(\nu_f = 0,\ 1,\ 2)$ states in N_2^+ as a function of the initial phase of the laser at time t_b. Field-dependent ionization is omitted for simplicity. For N equally weighted trajectories, the total probability of recollision

$$P_{tot} = \frac{1}{2N} \sum_{t_b} P(t_b) \tag{4.3}$$

is the average over the probabilities $P(t_b)$ for each trajectory. The extra factor of 1/2 is included instead of calculating trajectories on the increasing quarter-cycle that do not rescatter.

The total probability is about 3% for the conditions in Fig. 4.1. The cross section is zero below about 3.5 eV, so the probability drops sharply to zero below this energy in Fig. 4.1(b).

This model can be improved in a number of ways. For example, field-dependent ionization may lessen the contribution from electrons that return with low kinetic energy, which would increase the total probability. The Coulomb potential of the ion is omitted, but it can focus the recolliding electron wave packet and increase the inelastic scattering probability. Higher order returns also provide more opportunity for population transfer [[illegible]]. This enhancement could be significant, as Coulomb focusing and higher-order returns enhance correlated double ionization in helium by a factor of up to 5 [[illegible], [illegible]]. While additional effects that lower the probability may balance this enhancement, this discussion creates a reasonable expectation of some population transfer to the $B^2\Sigma_u^+$ state due to recollision.

4.3 Ellipticity and gain

Recollision can be efficient for an intense pump laser with linear polarization, but its efficiency decreases as the ellipticity of the pump laser polarization increases because the returning electron is displaced transversely from the parent ion due to the perpendicular field component. For example, a circularly polarized pump pulse enables the complete suppression of high harmonic generation and the detection of relatively weak emission lines, as shown in Figure 3.3(a). This polarization sensitivity also enables the polarization gating technique that generates attosecond pulses [[illegible]].

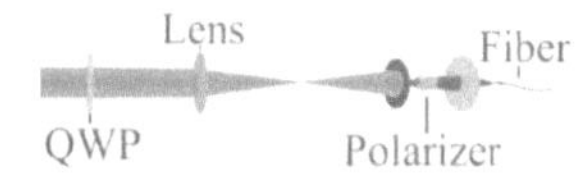

Figure 4.2: The setup to measure self-seeded gain in air, control the input laser polarization ellipticity, and characterize the output polarization.

The intensity of self-seeded N_2^+ emission decreases with increasing large ellipticity of the laser polarization, which is similar to the behaviour of high harmonic intensity in N_2. Since

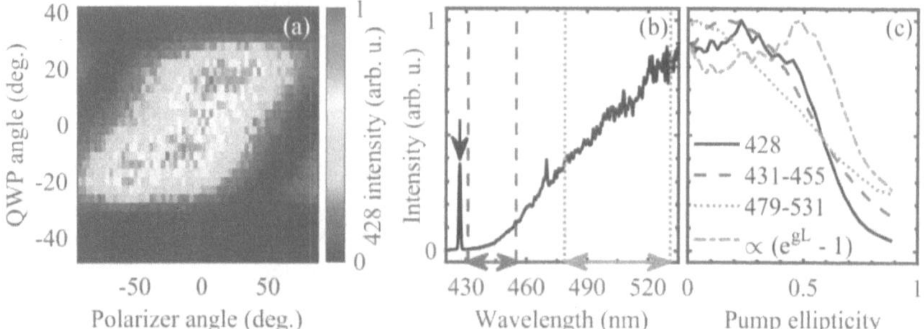

Figure 4.3: Air results. (a) The intensity of self-seeded gain at 428 nm (colour scale) as a function of the angle of the quarter wave plate (QWP) and polarizer. Zero QWP angle and zero polarizer angle correspond to the input linear polarization. Circular polarization is near ±45 degrees on the QWP axis. (b) The spectrum of the self-seeded emission at 428 nm. Three regions are indicated that are integrated in the next panel. (c) The integrated intensity of the peak at 428 nm and two regions of the self-seeding continuum as a function of the ellipticity of the laser polarization. The dot-dash line ($\propto e^{gL} - 1$) is gain accounting for the changing self-seed.

high harmonics are created during recollision, this inspired the suggestion that inelastic scattering during electron recollision could excite the population in the $B^2\Sigma_u^+$ state for N_2^+ gain [79]. The gas jet provides a unique opportunity to measure isolated gain and compare it directly with high harmonic generation. Before moving to the jet, we start by investigating the "air laser" in ambient air itself to better-understand how the ellipticity of the laser polarization influences self-seeded gain.

4.4 Air measurements

The air experiment is illustrated in Fig. 4.2. A spherical lens with a 30 cm focal length focuses an 800 nm ultrashort pulse (~25 fs, 3.3 mJ) to generate a plasma channel in air. A quarter wave plate (QWP) before the lens controls the ellipticity. The laser will generate its own probe pulse in air through spectral broadening during filamentation, so the orientation of the elliptical polarization does not matter in this case.

A polarizer and spectrometer monitor the emission after the focus. Figure 4.3(a) shows the intensity of self-seeded gain at 428 nm [$B^2\Sigma_u^+$ ($\nu = 0$) $\rightarrow$ $X^2\Sigma_g^+$ ($\nu = 1$)] in colour as a function of the QWP angle and the polarizer angle. The air measurement only considers gain at 428 nm, whereas most jet measurements only consider gain at 391 nm. We choose 428 nm in air because it is closer to the laser wavelength of 800 nm, so it is easier to generate self-seeding continuum.

Figure 4.3(b) is a sample spectrum showing the continuum and the self-seeded emission.

At each QWP angle in Fig. 4.3(a), the self-seeded emission reaches zero for some polarizer angles. As shown in Sec. 2.1.2, this corresponds to linearly polarized emission. Furthermore, the angle of the linear polarization rotates with the quarter wave plate, so the angle of the linear polarization follows the major axis of the input elliptical polarization.

Figure 4.3(a) shows a large variation of the self-seeded intensity as a function of the input polarization, but it does not account for the changing continuum self-seed. The self-seeding continuum also changes as a function of the input polarization [[illegible], [illegible]]. A downward arrow in Fig. 4.3(b) indicates the self-seeded emission, while vertical lines and horizontal double-arrows indicate two regions of continuum. The integrated intensities in these three regions behave differently as the ellipticity of the laser polarization increases in Fig. 4.3(c). The black curve ("428") is the intensity of the self-seeded emission, and it increases at small ellipticity before decreasing significantly at large ellipticity like Fig. 4.3(a). The nearby region of continuum ("431-455") behaves similarly. In contrast, the far region of continuum ("479-531") only decreases as ellipticity increases.

The gain-length product

$$gL = \ln \frac{I_{out}}{I_{in}} \tag{4.4}$$

adjusts the output intensity I_{out} to account for the input intensity I_{in}. If the length of the gain medium is constant, gL is proportional to gain. It is difficult to determine the intensity of the continuum under the peak at 428 nm in Fig. 4.3(b), but the nearby continuum can represent the input intensity.

The integrated intensity of the self-seeded emission

$$I_{out} = I_{in} + I_{gL} \tag{4.5}$$

includes both the self-seed I_{in} and the amplification I_{gL}. If the amplification is very large and the self-seed is negligible, like in Fig. 4.3(b), the output intensity is approximately the amplification ($I_{gL} \approx I_{out}$). If the intensity of the nearby continuum I_{cont} represents the self-seed ($I_{in} \propto I_{cont}$), the ratio of the output intensity given by Eq. 4.5 to the nearby continuum

$$\frac{I_{out}}{I_{cont}} \mathrel{\underset{\sim}{\propto}} e^{gL} - 1 \tag{4.6}$$

is related to gain. Figure 4.3(c) also shows this quantity. After removing part of the influence of self-seeding, gain [$\propto$ ($e^{gL} - 1$)] is flat as a function of the ellipticity of the laser polarization until about 0.5 where it increases a bit before decreasing rapidly.

This demonstrates that the variation of the continuum self-seed is partially responsible for the decrease of the self-seeded emission with ellipticity. These results show that self-seeded gain in air behaves in a way that is more complex than expected from recollision alone, but we must consider the uncontrolled propagation that modifies the laser pulse as it generates gain. This blurs the experimental conditions, which include the pump laser polarization [12]. The limited propagation distance in the jet offers more control and a direct comparison with high harmonics.

4.5 Jet measurements

In the absence of continuum self-seeding, a probe pulse is required to measure gain. Chapter 3 introduced an experimental setup that separates and frequency-doubles a portion of the 800 nm pulse to measure gain at 391 nm [$B^2\Sigma_u^+$ ($\nu = 0$) $\rightarrow$ $X^2\Sigma_g^+$ ($\nu = 0$)] in a gas jet. The jet measurement primarily studies gain at 391 nm, as it is close to the center wavelength of the second harmonic probe pulse, but we sometimes also seed 428 nm gain in the jet. Gain as a function of the delay between the pump and probe pulses reveals rotational, vibrational, and electronic dynamics. These dynamics are discussed more in Ch. 5 and 6.

The measurement in the jet is similar to air. Increasing the pump ellipticity will suppress recollision, as before. The polarization of the probe pulse is linear, which is particularly important due to molecular alignment from rotational wave packets that influences the transition strength [89, 112]. Figure 4.4(a) shows gain as a function of ellipticity of the pump pulse at three delays between the pump and probe pulses. The three delays are indicated by vertical lines on Fig. 4.4(b), which shows modulations from rotational wave packets in the $X^2\Sigma_g^+$ and $B^2\Sigma_u^+$ states that are superimposed on the decay of gain as a function of probe delay.

The shortest delay of 0.32 ps in Fig. 4.4(a) is near time overlap between the pump and probe pulses, which is like the air measurement. The results are similar to those in air; Gain is maximum at a small non-zero ellipticity, and it decreases at ellipticities larger than about 0.5. The latter two delays are during the early part of modulations from the half rotational revivals in Fig. 4.4(b), so they show the ellipticity dependence of gain during alignment and antialignment. At all three delays, the signature of recollision is missing from the ellipticity dependence.

Molecular alignment relative to the probe polarization in the $X^2\Sigma_g^+$ or $B^2\Sigma_u^+$ states will decrease or increase gain, respectively, as discussed in Sec. 5.1. Likewise, antialignment in either state has the opposite effect. Gain at the shortest delay in Fig. 4.4 increases when the pump polarization is parallel or perpendicular to the probe polarization. This is consistent with pump-induced impulsive alignment of both states at short delays. In contrast, the latter two delays only increase using either parallel or perpendicular polarization and may indicate the influence of alignment from only one state. It is noteworthy that gain exists using a circularly polarized pump pulse, and the circularly polarized pump pulse reduces the amplitude of the modulations.

4.5.1 Direct comparison with high harmonics

A process in N_2 that depends strongly on recollision should disappear as the ellipticity

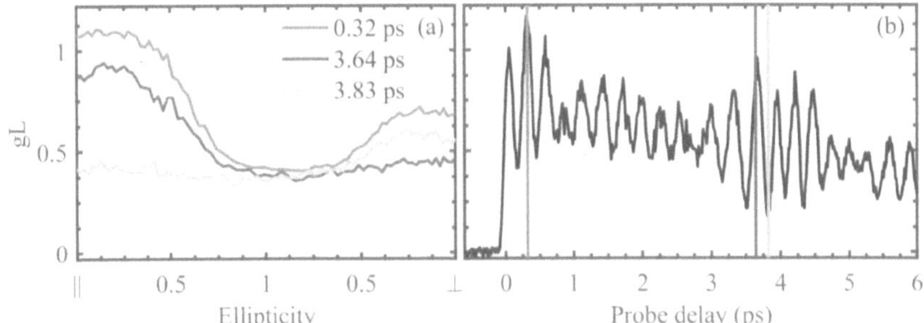

Figure 4.4: Gas jet results. (a) Gain at 391 nm as a function of the ellipticity of the pump pulse polarization at three probe delays. || and ⊥ correspond to linear pump polarization that is parallel or perpendicular to the linear probe polarization, respectively. $I_{Pump} = 7 \times 10^{14}$ W/cm^2 (b) Gain as a function of delay using a linearly polarized pump pulse with vertical lines indicating the probe delays in the left panel.

of the laser polarization increases, but gain in Fig. 4.4(a) and Fig. 4.3(c) does not. To demonstrate this, the gas jet enables a direct comparison of gain with high harmonic generation. In addition, broadening the spectrum of the probe pulse using self phase modulation extends gain in the jet to cover 391 nm and 428 nm simultaneously.

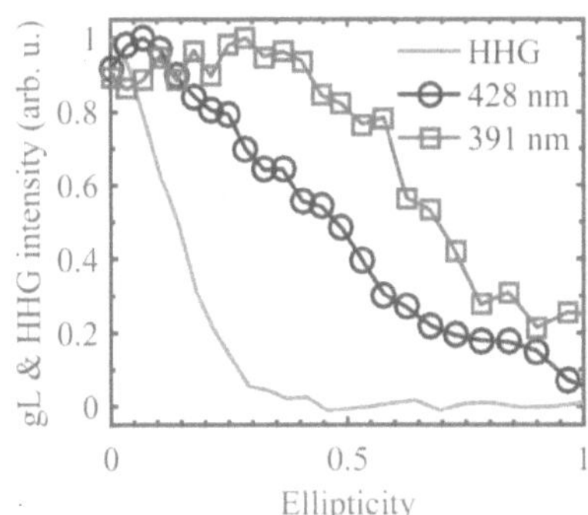

Figure 4.5: Comparison of gain with the intensity of high harmonics. The probe delay was small. $I_{Pump} = 4.5 \times 10^{14}$ W/cm^2

Figure 4.5 shows gain as a function of the ellipticity of the pump polarization. At 391 nm, the results essentially reproduce Fig. 4.4(a). The dependence of gain at 428 nm on the pump polarization is similar to 391 nm, except it reaches a maximum earlier and then decreases faster. The dynamics of gain may be different for the two transitions, but linear polarization is not essential for gain from either transition. Figure 4.5 also shows the intensity of high harmonics integrated over harmonic orders 11 to 21. High harmonics are rapidly suppressed as the ellipticity of the pump polarization increases, but gain persists without recollision like before.

4.6 Filamentation intensities

Filamentation is the anticipated delivery mechanism for an air laser built using N_2^+ gain. The traditional clamped intensity in a filament is about 5×10^{13} W/cm^2, which is lower than the pump intensities in Fig. 4.4 and Fig. 4.5. For an accurate comparison to filamentation results, we should use comparable intensities, but the traditional clamped intensity does not apply in all conditions. For example, the clamped intensity depends on the laser wavelength.

Intensity clamping arises from the balance between self-focusing and plasma defocusing during filamentation. Balancing the accumulated phase from Fig. 2.2 due to the Kerr effect and plasma

$$n_2 I = \frac{\rho}{2\rho_c} \tag{4.7}$$

yields an approximate value of the clamped intensity

$$I_c = \frac{\rho}{2n_2\rho_c} \tag{4.8}$$

for a fixed plasma density ρ, where $\rho_c = \epsilon_0 m_e \omega_0^2 / e^2$ is the critical plasma density and ω_0 is the laser frequency [19]. The clamped intensity

$$\begin{aligned} I_c &\propto \frac{\rho\lambda^2}{n_2} \\ &\propto \rho P_c \end{aligned} \tag{4.9}$$

is related to the critical power for self focusing P_c given by Eq. 2.19. The plasma density also depends on the laser intensity $[\rho = \rho(I)]$, so Eq. 4.9 is mostly qualitative.

The clamped intensity also depends on focusing geometry [20], and intensity spikes can exceed the clamped intensity by an order of magnitude [10]. Measurements and calculations of lasing in neutral nitrogen suggest a higher clamped intensity in air of 1.45×10^{14} W/cm^2 at 800 nm [19].

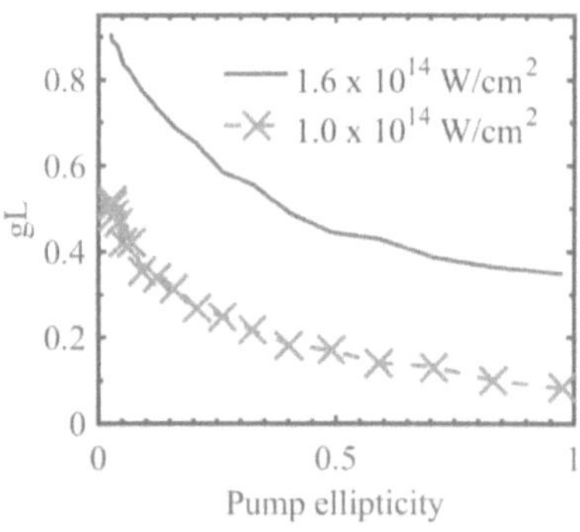

Figure 4.6: 391 nm gain as a function of the pump polarization ellipticity at two pump intensities (legend). The probe delay is near zero.

Figure 4.6 shows 391 nm gain as a function of the pump polarization ellipticity at two pump intensities that are comparable to the clamped intensity in a filament. Once again, gain persists at all pump polarizations.

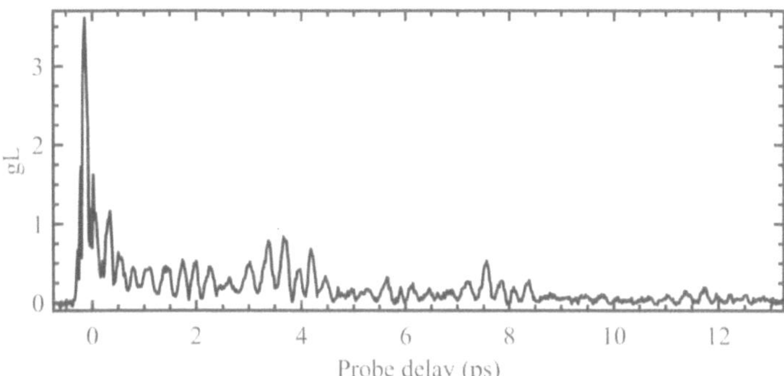

Figure 4.7: Gain as a function of probe delay in conditions comparable to Fig. 4.6 using a linearly polarized pump pulse.

Unlike Fig. 4.4 and 4.5, gain is maximum using linear pump polarization in Fig. 4.6. A new behaviour emerges at low intensity, which also behaves differently in time.

Multiple types of gain may be present in Fig 4.6. Figure 4.7 shows gain as a function of probe delay in similar conditions, and it looks different from Fig. 4.4(b). The usual decay of gain is replaced by the combination of a large spike near time overlap and a relatively small decay. The small decay looks similar to the usual exponential-like decay of gain. The probe pulse measures gain near zero probe delay in Fig 4.6, so gain includes both the large spike and the small decay. Self-seeding measurements that do not resolve dynamics would not be able to distinguish between the two regimes. Section 5.4 and 6.3 further explore the origin of the short-term gain.

+

The semiclassical model calculated the probability of inelastic excitation from the ground $X^2\Sigma_g^+$ to the excited $B^2\Sigma_u^+$ states during recollision by following the classical electron trajectories, approximating quantum diffusion, and using the cross sections for excitation. The model predicted that recollision will transfer a few percent of the population to the excited state, and it ignores some effects that may increase or decrease the predicted population transfer.

The experiments in air and the gas jet used the ellipticity of the ionizing laser polarization to prevent recollision. In all cases, we measured gain at large ellipticities where recollision is unlikely in N_2. The continuum self-seed in the air measurements also changed with laser polarization, so it was important to adjust for the varying self-seed intensity. In the jet, we made a direct comparison with high harmonic intensity and verified that gain persisted when recollision was suppressed. A new regime of short-term gain emerged during low-intensity measurements. The short-term gain was also available for all ellipticities.

Recollision may contribute to N_2^+ gain when the pump polarization is linear. In fact, the semiclassical calculation suggests that it should. However, these experimental results using elliptical polarization show that recollision is not essential for N_2^+ gain [3]. This was verified [109].

Notably, the pump pulse coupling between the $X^2\Sigma_g^+$ and $A^2\Pi_u$ states around 800 nm also contributes to the polarization dependence of gain. For a linearly polarized pump pulse, the initial alignment of the $X^2\Sigma_g^+$ state relative to the polarization direction can prevent these perpendicular transitions from depleting its population. Therefore, a laser pulse with elliptical or time-dependent polarization can optimize gain by tuning the strength of these transitions using molecular alignment [106, 110]. Chapter 6 will show that a second separate excitation pulse can control emission in addition to gain using this coupling, but Ch 5 first explores the decay and dynamics of unperturbed gain.

Chapter 5

Gain dynamics

5.1 Dynamics

Chapter 3 introduced a measurement of N_2^+ gain using a narrow gas jet to limit propagation distance, and Ch. 4 applied the measurement at fixed probe delays to show that recollision is not a prerequisite for gain. The time evolution of gain provides information about additional electronic, vibrational, and rotational dynamics [[illegible]]. Figure 5.1 shows gain as a function of probe delay at four gas jet positions along the gas jet flow direction. It is similar to both Fig. 4.4(b) and Fig. 3.16. Modulations due to rotational wave packets are superimposed on an exponential-like decay of gain.

The gas jet position relative to the laser focus is an important parameter in the experiment because it controls the length, density, and temperature of the gain medium, and the size and curvature of the lasers. Figure 5.1 shows that the gas jet position influences the total amount of gain, the modulations superimposed on the decay of gain, and the rate of the decay of gain. $x_0 \approx 250\ \mu m$ is the shortest distance between the nozzle and the focus without obscuring the laser based on the focusing geometry and the size of the nozzle enclosure.

At the position x_0 in Fig. 5.1, gain decays rapidly and the duration of the modulations is relatively short. The maximum gain increases and then decreases as the gas jet position increases. Due to the expansion of the jet, density is lower and the medium length L is longer at farther positions. Density influences the number inverted molecules and gain g. At far jet positions, the density should decrease the gain length product gL while the length increases it. The length increases due to expansion in one dimension along the laser propagation direction, but the density falls at a faster rate due to expansion into multiple spatial dimensions. Therefore, the optimum gain at $x_0 + 250\ \mu m$ in Fig. 5.1 may indicate an optimum density for generating inversion because the gain length product should otherwise fall monotonically as the jet position increases. Plasma defocusing from ionization at higher densities may alternatively reduce gain at the closest position. The gain decay is slower at farther jet positions in Fig. 5.1, but the fastest decay is distinct from short-term gain (*e.g.* Fig. 4.7) because it is an exponential-like decay instead of a single spike at zero probe delay.

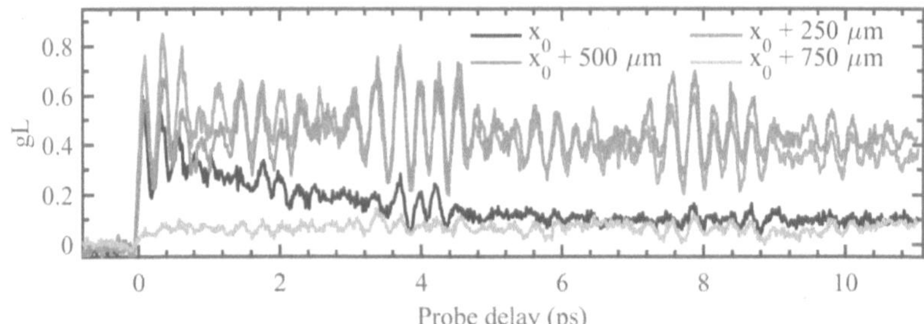

Figure 5.1: Gain as a function of probe delay at four positions of the gas jet along the gas flow direction. x_0 is the closest position without obstructing the laser. $I_{Pump} = 4 \times 10^{14}$ W/cm^2

The modulations superimposed on the decay of gain are due to rotational wave packets in the states of the ion. These wave packets begin prior to ionization in the neutral molecule. Impulsive excitation can occur when the duration of the excitation pulse is much shorter than the response time of the system, which requires that the bandwidth of the pulse exceeds the spacing of the energy levels. Under these conditions, stimulated Raman scattering among the states within the bandwidth of the pulse can establish a coherent superposition known as a wave packet, as illustrated in Fig. 5.2. Due to the close spacing of rotational states relative to the bandwidth of ultrashort pulses, a rotational wave packet can form in the ground state of nitrogen.

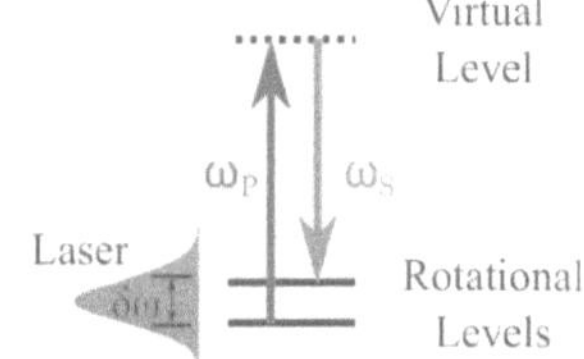

Figure 5.2: A cartoon of impulsive Raman excitation. If the duration of the pulse is shorter than the response of the system, impulsive excitation can establish a wave packet by stimulated Raman scattering within the bandwidth of the pulse.

In the rigid rotor approximation, the internuclear distance is fixed and the rotational constant B_0 describes the rotation of the molecule. This omits the centrifugal distortion constant D_0 that can be an important particularly for high rotational levels. The energies of the rotational states

$$\epsilon_J = B_0 J(J+1) \tag{5.1}$$

depend on the rotational constant and the rotational quantum number J. Impulsive excitation can form a coherent superposition of these states

$$|\Psi(t)\rangle = \sum_J A_{J,M} e^{-i\epsilon_J t} |J, M\rangle \tag{5.2}$$

that proceeds to evolve field-free once the laser has passed, where $A_{J,M}$ is the coefficient of a particular state $|J, M\rangle$ and M is the magnetic quantum number. The selection rules for an electric dipole transition allow stimulated Raman scattering to couple each rotational state J to the second closest $J + 2$, so odd and even J states do not couple due to the symmetry of the potential. Including the rotational energies in the superposition

$$|\Psi(t)\rangle = \sum_J A_{J,M} e^{-iB_0 J(J+1)t} |J, M\rangle \tag{5.3}$$

reveals that the system returns to its initial condition after $T_{rev} = \pi/B_0$, which is the full rotational revival. The half revival occurs at $T_{rev}/2$ when the system returns to its initial condition with a π phase shift, and the quarter revival occurs at $T_{rev}/4$ when the system returns to its initial condition but split into two copies that differ in phase by $\pi/2$.

Rotational wave packets modulate the average alignment of the molecules. The average value of $\cos^2\theta$ is a measure of molecular alignment that is 1 for aligned molecules, 0 for antialigned molecules, and 1/3 for a uniform distribution. The average alignment

$$\begin{aligned} \langle \cos^2\theta \rangle =& \langle \Psi(t)| \cos^2\theta |\Psi(t)\rangle \\ =& \sum_J \big(|A_{J,M}|^2 C_{J,J,M} \\ &+ |A_{J,M}||A_{J+2,M}| \cos\left[(\epsilon_{J+2} - \epsilon_J)t + \phi_{J,J+2}\right] C_{J,J+2,M} \big) \end{aligned} \tag{5.4}$$

also returns to the initial condition at the full revival time and reverses at the half revival time, where $C_{J,J',M} = \langle J, M| \cos^2\theta |J', M\rangle$ and $\phi_{J,J'}$ is the initial phase difference between $|J, M\rangle$ and $|J', M\rangle$ [[illegible]].

The pump pulse can form a rotational wave packet in neutral nitrogen that will modulate the index of refraction and birefringence of the medium [[illegible]]. Sec. 5.3.3 considers how this time dependent index of refraction modulates the phase of a pulse during propagation. The initial temperature of the medium determines the rotational population distribution before the pump pulse arrives, and much of the nitrogen will ionize near the peak of the pulse transferring its rotational wave packet to the states of the ion; however, it does this non-uniformly in part because the ionization depends on the angle between the molecular axis and the electric field [[illegible], [illegible]]. After ionization, impulsive excitation can continue in the states of N_2^+ that evolve on different timescales. This complicated sequence creates rotational wave packets that contain a lot of information but are difficult to model.

After the pump pulse passes, the rotational wave packets modulate molecular alignment in the $B^2\Sigma_u^+$ and $X^2\Sigma_g^+$ states. Gain is sensitive to the alignment in both of these states because the probe pulse $\vec{E}_{Probe}$ is linearly polarized and the transition dipole moment $\vec{d}_{BX}$ is parallel, so the transition is maximum when molecules in the $B^2\Sigma_u^+$ state are aligned with the laser polarization. Absorption is equivalently minimum when molecules in the $X^2\Sigma_g^+$ are antialigned with the laser polarization. Therefore, gain and absorption on transitions between these electronic states include their alignment signal [[illegible], [illegible]]. Section 5.3.2 uses the alignment signal to determine contribution from the rotational wave packet in each state.

5.1.1 Lasing without inversion

The sensitivity of gain to the alignment in the $X^2\Sigma_g^+$ and $B^2\Sigma_u^+$ states creates a possibility for lasing without inversion [89]. In a simple example, absorption can be completely suppressed if the all of the molecules in the $X^2\Sigma_g^+$ state are antialigned, and then any non-antialigned molecules in the $B^2\Sigma_u^+$ state would generate net gain.

Equivalently, an individual rotational state in one electronic state may couple to any rotational state in the other electronic state, but the coherent superposition of two (or more) rotational states in the other electronic state can destructively interfere to cancel the transition amplitude to the individual state [145, 146, 147, 148, 149, 150]. This can happen in N_2^+ for absorption to the rotational states in $B^2\Sigma_u^+$ state due to the rotational wave packet in the $X^2\Sigma_g^+$ state and for emission to the rotational states in the $X^2\Sigma_g^+$ state due to the rotational wave packet in the $B^2\Sigma_u^+$ state.

5.2 Exponential-like decay of gain

The decay of gain occurs over tens of picoseconds but depends on the gas jet position. The population of a collection of isolated N_2^+ ions would remain constant until spontaneously radiatively decaying, but the real environment introduces quenching mechanisms that hasten the decay of population. For example, the gain medium contains energetic electrons that can collide with ions to redistribute the population [151]. The large inelastic scattering cross section for the $B^2\Sigma_u^+$ to $X^2\Sigma_g^+$ transitions motivated the exploration of excitation during recollision in Ch. 4, but free electrons in the plasma can also scatter from ions after ionization.

The rate $R_\sigma = 1/\tau_\sigma$ of electron-ion inelastic collisions is the inverse of the time between collisions τ_σ. The electrons in the plasma have a distribution of energies, and the energy of an electron corresponds to an average speed v. An electron traveling at speed v must occupy the volume per unit time σv to inelastically scatter from an ion, where σ is the cross section that depends on the electron speed. The rate of excitation

$$R_\sigma = \rho\sigma v, \tag{5.5}$$

or the number of electrons occupying that volume per unit time, and the time between collisions

$$\tau_\sigma = \frac{1}{\rho\sigma v}, \tag{5.6}$$

simply require the number per unit volume ρ.

If inelastic scattering depopulates the $B^2\Sigma_u^+$ state, Eq. 5.5 shows that the rate should decrease with density. The decay is slower at farther jet positions in Figure 5.1. Density

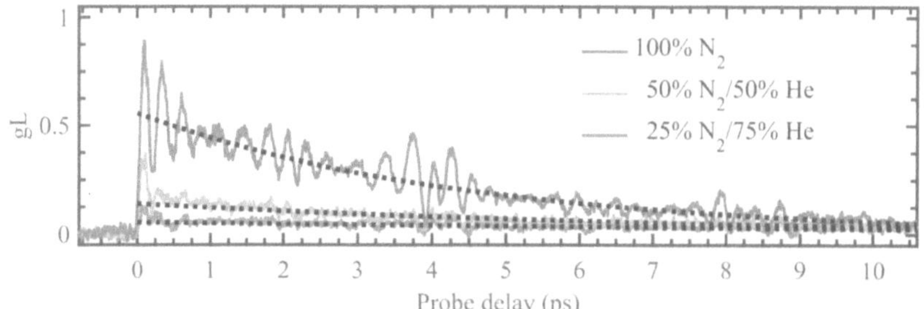

Figure 5.3: Gain as a function of probe delay using three mixtures of nitrogen and helium. $I_{Pump} = 3 \times 10^{14}$ W/cm^2

decreases as the gas jet position increases, so the data is consistent with collisional quenching. For the $B^2\Sigma_u^+$ to $X^2\Sigma_g^+$ transitions in N_2^+, the cross-section peaks at approximately 3×10^{-16} cm^2 near 3.2 eV. Assuming an average electron energy of 5 eV and a density of 2.5×10^{18} cm^{-3}, the estimated time between collisions $\tau_\sigma \approx 15$ ps is consistent with the decays in Fig. 5.1.

This process is different from radiative decay because the populations of the two states equalize instead of decaying to the lower energy state. Ignoring other effects, the set of equations

$$\begin{aligned} \frac{dN_B}{dt} &= R_\sigma N_X \\ \frac{dN_X}{dt} &= R_\sigma N_B. \end{aligned} \tag{5.7}$$

determines the evolution of the populations N_X and N_B of the $X^2\Sigma_g^+$ and $B^2\Sigma_u^+$ states, respectively. The population inversion

$$\Delta N = N_B - N_X \tag{5.8}$$

is the difference of the two populations. It evolves according to the equation

$$\begin{aligned} \frac{d\Delta N}{dt} &= -R_\sigma(N_B - N_X) \\ &= -R_\sigma \Delta N. \end{aligned} \tag{5.9}$$

Integrating the time dependence of the population inversion yields an exponential decay

$$\Delta N(t) = C_1 \exp\{-R_\sigma t\} + C_2, \tag{5.10}$$

where C_1 and C_2 are constants. Assuming an arbitrary initial inversion $C_1 = \Delta N_0$, and that all of the population is influenced by the decay mechanism ($C_2 = 0$), the population inversion

$$\Delta N(t) = \Delta N_0 \exp\{-R_\sigma t\}. \tag{5.11}$$

decays exponentially in time. This is consistent with the exponential-like decay of gain that we observe. Furthermore, Eq. 5.7 shows that inelastic scattering continues to incoherently mix the states after the population inversion decays.

Equation 5.5 suggests that the density of electrons and ions can control the decay. The backing pressure of the gas jet influences the density, but it also influences other properties of the jet. Gas mixtures can modify the density in a controlled way without changing the jet much. Figure 5.3 shows gain as a function of probe delay using three different mixtures of nitrogen and helium. Helium has a high ionization potential, so the number of ions and electrons scales with the partial pressure of nitrogen. As the nitrogen concentration decreases, the rate of decay is slower.

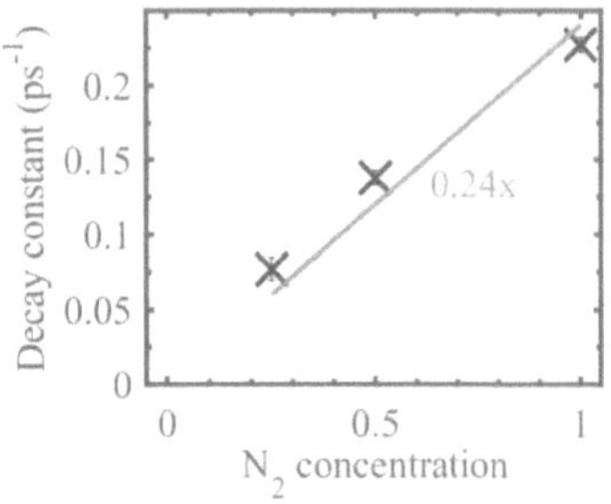

Figure 5.4: The exponential decay constant from the fits in Fig. 5.3 as a function of nitrogen concentration. The solid line and equation correspond to a linear least-squares fit through zero where x is the concentration.

A least-squares fit using an exponential decay model is superimposed on the each decay in Fig. 5.3 (black dotted lines), and they quantify the amplitudes and rates of the decays. The amplitudes extracted from the three fits scale quadratically with concentration. The quadratic scaling of gain was previously attributed to the collective nature of the emission [67]. Surprisingly, the variation of gain is not as dramatic in Fig. 5.1 at jet positions between x_0 to $x_0 + 500$ μm, but gain falls dramatically at $x_0 + 750$ μm. This suggests that the quadratic scaling may not continue at high density. As an alternative explanation for the quadratic scaling, stochastic many body effects could also contribute to quadratic scaling during the generation of gain, but this possibility has not been explored.

Figure 5.4 shows the decay constants from the exponential decay fits superimposed on Fig. 5.3. The decay constants extracted from the three fits scale linearly with concentration, as predicted by Eq. 5.5. These results suggest that inelastic scattering equalizes the population of the $B^2\Sigma_u^+$ and $X^2\Sigma_g^+$ states.

At longer delays, a single exponential decay fit is sometimes insufficient. This indicates that Eq. 5.11 is missing something. For example, a more general form of Eq. 5.6

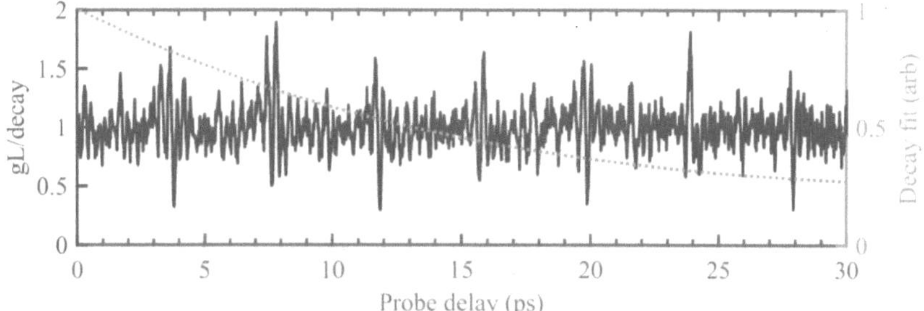

Figure 5.5: Gain as a function of probe delay factored into two components: the underlying exponential decay of gain (right-hand axis), and the superimposed modulations. The modulations are divided by the underlying decay to remove its influence (left-hand axis). $I_{Pump} = 2 \times 10^{14}$ W/cm^2

$$R_\sigma = \int_0^\infty \rho_0(\epsilon)\sigma(\epsilon)v(\epsilon)\mathrm{d}\epsilon \tag{5.12}$$

accounts for the energy-dependence of the electrons in the plasma, where ϵ is the electron energy, $\sigma(\epsilon)$ is the energy-dependent cross-section, $v(\epsilon)$ is the electron speed, and $\rho_0(\epsilon)$ is the electron energy distribution

$$\int_0^\infty \rho_0(\epsilon)\mathrm{d}\epsilon = \rho. \tag{5.13}$$

This correction would increase the precision of the model independent of delay.

The energy distribution of electrons changes as a function of time as they exchange energy with the ions. Therefore, the electron energy distribution $\rho_0(\epsilon, t)$ is a function of time. In that case, the rate of mixing

$$R_\sigma(t) = \int_0^\infty \rho_0(\epsilon, t)\sigma(\epsilon)v(\epsilon)d\epsilon. \tag{5.14}$$

also depends on time. This dynamic mixing rate may capture the non-exponential heating or cooling of electrons during the decay of gain.

The dynamic mixing rate may be particularly important in air filamentation. The initial rate of collisional state mixing should be lower in air due to the lower initial electron temperature [ ], but thermalization with other degrees of freedom is particularly important due to the abundance of neutral molecules. Neutral molecules may provide a sink for electron energy, which a dynamic mixing rate including many more states could capture.

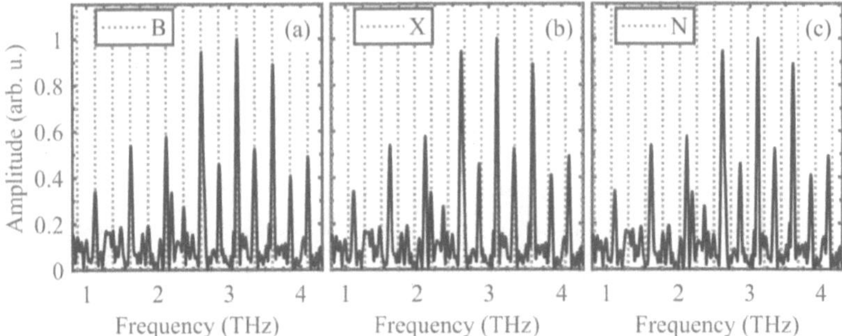

Figure 5.6: The Fourier transform of the modulations in Fig. 5.5 and the rotational beat frequencies of the (a) $B^2\Sigma_u^+$ and (b) $X^2\Sigma_g^+$ states of N_2^+ and (c) the ground state of N_2.

5.3 Modulations from rotational wave packets

5.3.1 Rotational dephasing

As the gain decays in Fig. 5.1 and Fig. 5.3, the amplitude of the superimposed modulations also decays. This indicates dephasing of the rotational wave packets in the $B^2\Sigma_u^+$ and $X^2\Sigma_g^+$ states that modulate molecular alignment. Figure 5.5 shows gain in an unusual way: the exponential decay is separated from the modulations. The right-hand axis shows the result of a least-squares fit on the data using an exponential decay model. The left-hand axis shows gain divided by the fit, which scales the modulations to the decaying gain. The amplitude of the modulations after correcting for the decay is roughly constant, so the modulations and the gain both decay at the same rate.

Rotational wave packets can decay due to decoherence, and inelastic scattering is an incoherent process, so collisional state mixing may also be responsible for the decay of rotational coherence. Inelastic scattering mixes the $B^2\Sigma_u^+$ and $X^2\Sigma_g^+$ states, which also mixes the rotational wave packets. The rotational wave packets evolve on different timescales in each state, so they drift out of phase by spending time in the other state. The constant stochastic mixing of the two rotational wave packets as they evolve is a source of rotational decoherence. Equation 5.7 describes the mixing, but it ignores rotational states. The rate of state mixing R_σ is consistent with the decay of both rotational wave packets and gain.

5.3.2 Rotational wave packets

The alignment signal is captured by the modulations of gain, and it provides information about the rotational wave packets created by the pump pulse [112]. The beating between two adjacent rotational states J and $J+2$ in the rotational wave packet modulates the average alignment in Eq. 5.4 with a frequency corresponding to their difference in energy

$$
\begin{aligned}
\Delta\epsilon &= \epsilon_{J+2} - \epsilon_J \\
&= B_0(4J + 6).
\end{aligned}
\tag{5.15}
$$

Table 5.1 lists the rotational constants for the ground state of neutral nitrogen (N) and the $X^2\Sigma_g^+$, $A^2\Pi_u$, and $B^2\Sigma_u^+$ states of the cation [[illegible], [illegible]]. Figure 5.6 shows the Fourier transform of the modulations in Fig. 5.5, which extend to 39 ps. The positive half-cycle of a sinusoidal function scales the data so it smoothly decreases to zero, and zero-padding in the time domain interpolates the data in the frequency domain.

The three panels of Fig. 5.6 include vertical lines at the rotational beat frequencies for (a) the $B^2\Sigma_u^+$ state, (b) the $X^2\Sigma_g^+$ state, and (c) neutral nitrogen. The agreement between the peaks and the expected frequencies is excellent for the $B^2\Sigma_u^+$ state, while a few peaks can be assigned to the $X^2\Sigma_g^+$ state. Therefore, the rotational wave packet in the $X^2\Sigma_g^+$ state is weak in this experiment. Neutral nitrogen does not modulate the transition strength, and its frequencies are also correspondingly weak.

Table 5.1: B_0 for the ground state of nitrogen (N) and three states of N_2^+.

State	B_0 (cm^{-1})
N	1.99
X	1.93
A	1.74
B	2.07

A large inversion may contribute to the relative weakness of the rotational wave packet in the $X^2\Sigma_g^+$ state, as molecular alignment can only influence population transfer from the $X^2\Sigma_g^+$ state when the state is populated. Faster rotational dephasing in the $X^2\Sigma_g^+$ state is another explanation, which Sec. 6.2.2 explores.

The temperature of the gas jet determines the distribution of initial rotational populations. As the jet expands, Fig. 5.1 shows wider modulations in time. The jet cools as it expands, and a narrower initial distribution of rotational population in N_2 may become a narrower final distribution in the states of N_2^+. The resulting narrower distribution of rotational beat frequencies is consistent with wider modulations in time at farther jet positions.

The final distributions of rotational populations may be different in the $B^2\Sigma_u^+$ and $X^2\Sigma_g^+$ states. This could enable partial inversion among some rotational states [[illegible], [illegible], [illegible]]. As discussed previously, the rotational coherence also modulates the strength of each transition and can generate transient gain [[illegible], [illegible], [illegible], [illegible], [illegible]]. In addition, the amplified emission propagates through a phase modulator due to the rotational wave packets [[illegible]]. This molecular modulator may also influence gain [[illegible]].

5.3.3 The molecular modulator

The structure of the modulations looks more complex than alignment modulations from rotational wave packets during high harmonic generation in N_2 [[illegible]]. The low temperature in the jet generates long-lasting rotational revivals, and rotational excitation occurs before, during, and after ionization. Furthermore, the evolving molecular alignment corresponds to a time dependent refractive index. The time dependent refractive index due to the Kerr effect introduced new frequencies in Eq. 2.13 to 2.15, so the molecular modulator may also introduce new frequencies to the emission. A simple model can illustrate this effect.

The nonlinear Schrödinger equation

$$\frac{\partial E}{\partial z} = i\frac{\beta_2}{2}\frac{\partial^2 E}{\partial t^2} - i\gamma|E|^2 E \tag{5.16}$$

captures self-phase modulation and dispersion of the slowly varying electric field envelope E, where γ is a nonlinearity parameter that is proportional to the nonlinear refractive index and β_2 is the second-order propagation constant.

The split-step Fourier method is a computational approach to solving Eq. 5.16 by applying frequency and time dependent phase in their respective domains over a short propagation distance Δz [ ]. For an initial envelope $E_i(t)$ in time, the Fourier transform yields the spectrum $E_i(\omega)$. The envelope after a partial step

$$E_d(\omega) = E_i(\omega)\exp\{-i\beta_2\omega^2\Delta z/2\} \tag{5.17}$$

includes the spectral phase from dispersion. The inverse Fourier transform yields the envelope in time after the partial step $E_d(t)$. The envelope after the full step

$$E_f(t) = E_d(t)\exp\{-i\gamma|E_d(t)|^2\Delta z\} \tag{5.18}$$

includes the phase due to the nonlinearity applied in the time domain. The next propagation step proceeds using the final envelope from the previous step as the input envelope. Figure 2.1 shows the results of this calculation without a molecular modulator.

The molecular modulator introduces a time dependent phase like self-phase modulation. A simple molecular modulator

$$n = n_0 + \Delta n\cos(2\pi t_M/T_M) \tag{5.19}$$

varies sinusoidally in time, where Δn and T_M are the amplitude and period of the modulator, respectively. If the pump pulse creates the molecular modulator, it moves at the group velocity of the pump pulse and the time in the modulator frame

$$t_M = t + \frac{z}{v_{g,probe}} - \frac{z}{v_{g,pump}}, \tag{5.20}$$

depends on the group velocity of the probe $v_{g,probe}$ and pump $v_{g,pump}$.

The corresponding time-dependent phase

$$\Delta\Phi_{MM}(t) = \frac{2\pi\Delta n}{\lambda}\cos(2\pi t_M/T_M)\Delta z \tag{5.21}$$

of the molecular modulator is like the time dependent phase due to the Kerr effect

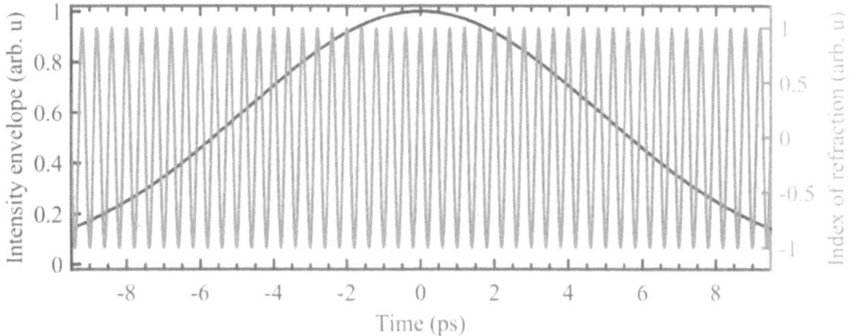

Figure 5.7: The intensity envelope of a laser pulse (left) and a modulated index of refraction (right) that illustrate the molecular modulator.

$$\begin{aligned} \Delta\Phi_{SPM}(t) &= \frac{2\pi n_2 I(t)}{\lambda}\Delta z \\ &= \gamma |E_d(t)|^2 \Delta z, \end{aligned} \tag{5.22}$$

except the phase of the molecular modulator does not depend on intensity. In this case, the envelope after the full step

$$E_f(t) = E_d(t)\exp\{-i\left[\Delta\Phi_{SPM}(t) + \Delta\Phi_{MM}(t)\right]\} \tag{5.23}$$

can include the phase of both effects.

Figure 5.7(a) shows the intensity of an initially Gaussian pulse as a function of time and a simple molecular modulator, where the pulse duration $\tau_F = 8$ ps and modulator period $T_M = 400$ fs are comparable to the duration of the free induction decay and the period of the alignment modulations during N_2^+ lasing, respectively. The amplitude of the molecular modulator $\Delta n = 6 \times 10^{-6}$ comes from the difference between the parallel and perpendicular static polarizabilities of neutral nitrogen [[illegible]].

The ratio of the modulator period to the pulse duration

$$\frac{T_M}{\tau_F} \tag{5.24}$$

strongly influences the results. The period of the modulator is much shorter than the pulse duration in Fig. 5.7, and the modulator converts the initial pulse to sidebands as a function of propagation distance in Figure 5.8(a). Figure 5.8(b) shows the intensity of the input and the first and second sidebands as a function of propagation distance. The intensities of the sidebands rise and fall as they exchange energy.

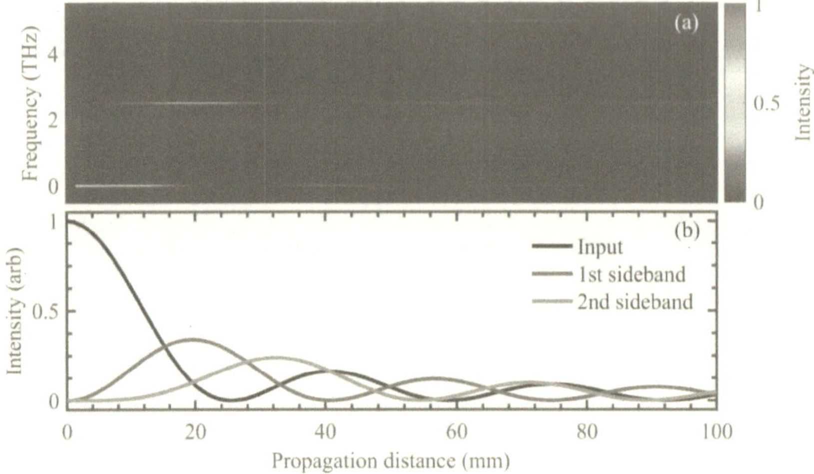

Figure 5.8: (a) The intensity of the laser pulse (colour) propagating in the presence of the index of refraction in Fig. 5.7 as a function of frequency and propagation distance. (b) The intensity of the center frequency of the input and the 1st and 2nd sidebands as a function of propagation distance. $n_2 = 0$, $\beta_2 = 0$, $\Delta z = 10\ \mu$m, $\lambda = 400$ nm, $\tau_F = 8$ ps, $T_M = 400$ fs, and $\Delta n = 6 \times 10^{-6}$

The molecular modulator in the N_2^+ gain medium contains a distribution of rotational beat frequencies for the states of the ion and the neutral molecule. Unlike self-phase modulation and other nonlinear optical effects, Eq. 5.21 shows that the molecular modulator phase does not depend on the intensity of the emission. Therefore, even low intensity lasing emission may develop new temporal and spectral features as it propagates, so our experimental technique of using a weak probe pulse and short propagation distance may not eliminate the effects of the molecular modulator. In addition, this effect could influence N_2^+ gain if sidebands overlap transitions involving different rotational states. The period of the modulator determines the sideband spacing, so the rotational beat frequencies may naturally create sidebands that overlap other transitions. This illustrates one influence of propagation on N_2^+ lasing [91].

5.4 Short-term gain

The exponential-like decay diminished while studying the role of recollision near the clamped intensity of the filament in Sec. 4.6, and short-term gain emerged instead. Figure 5.9(a) shows the exponential-like decay and short-term gain at both 391 nm and 428 nm, and the short-term gain has no weak underlying decay unlike Fig. 4.7. Figure 5.9(b) shows that

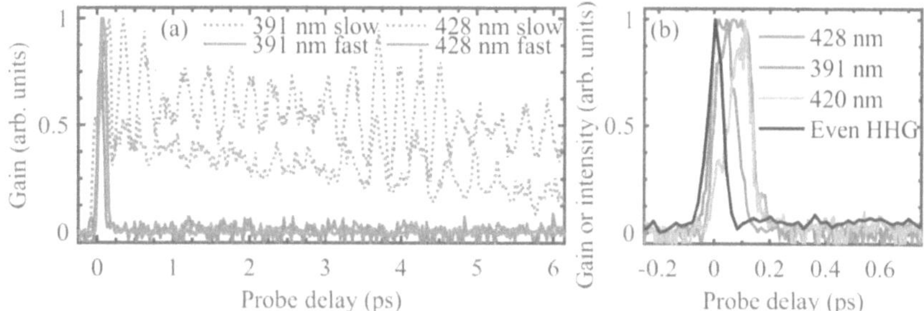

Figure 5.9: (a) Two timescales of gain using different gas jet positions and laser intensities. Short-term gain (fast) and the exponential decay (slow) appear at both 391 nm and 428 nm. (b) Comparison of short-term gain with the intensity of even-order high harmonics. Short-term gain also appears at 420 nm.

short-term gain exists during and immediately after zero probe delay, which is indicated by the intensity of even harmonics.

Short-term gain also appears at 420 nm [$B^2\Sigma_u^+$ ($\nu = 2$) $\rightarrow$ $X^2\Sigma_g^+$ ($\nu = 3$)] and 424 nm [$B^2\Sigma_u^+$ ($\nu = 1$) $\rightarrow$ $X^2\Sigma_g^+$ ($\nu = 2$)]. They behave similarly, so Fig. 5.9(b) only shows gain at 420 nm. Gain at these transitions was reported once previously without a separate probe pulse [ ]. With a probe pulse, we observe only short-term gain at these transitions.

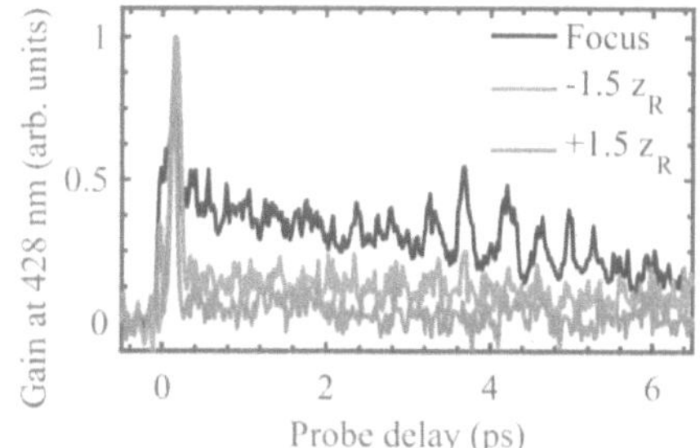

Figure 5.10: Gain as a function of probe delay at three gas jet positions along the laser propagation direction. The intensity of the pump pulse decreases by a factor of ~3.2 at ~$1.5 z_R$. $I_{Pump} = 2 \times 10^{14}$ W/cm^2.

Short-term gain generally appears at low pump intensity. The gas jet position along the laser propagation direction controls the intensity of the pulses in Figure 5.10, which avoids moving optics to change the energy of the beam. Short-term gain at 428 nm emerges at lower pump intensity on both sides of the focus, but an exponential-decay exists in the focus. Short-term gain is hard to distinguish from the exponential-like decay of gain in self-seeded experiments without a probe pulse because the time dependence is absent.

The short duration of this gain suggests that it relies on wave-mixing between the pump and probe pulses. One possibility is Raman gain in a V-system involving the $X^2\Sigma_g^+$, $A^2\Pi_u$, and $B^2\Sigma_u^+$ states. If the population of the $B^2\Sigma_u^+$ state exceeds the $A^2\Pi_u$ state, absorption of pump photons on $A^2\Pi_u$ to $X^2\Sigma_g^+$ transitions around 800 nm is accompanied by the stimulated emission of photons on $B^2\Sigma_u^+$ to $X^2\Sigma_g^+$ transitions [158]. This can occur even when the population of the $B^2\Sigma_u^+$ state is less than the population of the $X^2\Sigma_g^+$ state, which may occur at low pump intensity. Section 6.3 further explores short-term gain using an additional non-ionizing 800 nm control pulse.

+

The decay of gain is consistent with the decay of population inversion due to state mixing by collisional quenching that equalizes population in the $B^2\Sigma_u^+$ and $X^2\Sigma_g^+$ states. The concentration of nitrogen controlled the electron and ion density, and the decay changed according to the model of inelastic scattering. The resulting redistribution of electron energy may be an important correction.

The modulations superimposed on the gain decay are due to rotational wave packets in the $B^2\Sigma_u^+$ and $X^2\Sigma_g^+$ states that modulate molecular alignment. The Fourier spectrum of the modulations in the alignment signal showed a better match to the rotational beat frequencies for the $B^2\Sigma_u^+$ state, which suggests a large population inversion or fast rotational decoherence in the $X^2\Sigma_g^+$ state. Inelastic collisions randomly mix the rotational wave packets as they evolve, which is consistent with rotational dephasing on the timescale of the exponential-like decay. The rotational wave packets create a molecular modulator that influences emission and gain.

In addition to the usual exponential-like decay of gain, short-term gain is consistent with Raman gain in a V-system formed from the $X^2\Sigma_g^+$, $A^2\Pi_u$, and $B^2\Sigma_u^+$ states. Short-term gain was separated from the exponential-like decay at low pump intensity, but the two timescales are hard to distinguish without a probe.

The gas jet limits propagation, but the generation of gain, the evolution of gain, the probe interaction, and the molecular modulator still matter. Chapter 6 introduces a non-ionizing control pulse that modifies gain and emission. When the pump pulse is relegated to creating the gain medium, the control and probe pulses perform time resolved pump-probe spectroscopy on the ion itself.

Chapter 6

Control of gain and emission

6.1 Pumping isolated N_2^+

Chapter 5 discussed time-resolved pump-probe measurements of the gain in N_2^+. The pump pulse is multifaceted in these traditional measurements because it not only ionizes the neutral molecules, but it also excites both the neutral molecules prior to ionization and the ions after ionization. This is reflected in the rotational and vibrational dynamics. For example, impulsive alignment of the neutral molecule and angle-dependent ionization influence the excitation of rotational wave packets in the states of the ion, which increases their complexity. Population exchange between the $X^2\Sigma_g^+$ and $A^2\Pi_u$ states depletes the ground state population using common near-infrared pump wavelengths only after ionization, and the resulting vibrational motion in the $A^2\Pi_u$ state can temporarily trap population.

This chapter uses an additional non-ionizing excitation pulse that can initiate dynamics with more control after the pump pulse generates the gain medium. Figure 6.1(a) illustrates these three collinear pulses. In this experiment, the spectrum of the additional 800 nm control pulse is the same as the pump pulse. The polarization of the control pulse is nearly circular, while the pump and probe polarizations remain linear and parallel. Figure 6.1(b) shows gain as a function of both probe and control delays, and Fig. 6.1(c) shows horizontal lineouts like the traditional measurement. The following three sections consider different features of these figures.

At negative control delays, the control pulse only interacts with neutral nitrogen. While it can align N_2 prior to ionization, it is otherwise similar to the usual case without a control pulse [118]. Therefore, the negative control delay −0.5 ps in Fig. 6.1(c) looks like the results of traditional measurements. At positive control delays, gain depends strongly on the delay. The diagonal dashed line superimposed on Fig. 6.1(b) indicates overlap between the control and probe pulses, and it marks an abrupt change of the gain where the order of the control and probe pulses change. Figure 6.2 highlights this separation between Pump–Probe–Control and Pump–Control–Probe measurements at the control delay of 7 ps. Both cases differ from the traditional measurement.

The most prominent feature of Fig. 6.1 and 6.2 is decreased gain in the Pump–Probe–Control measurements. Section 6.1.3 will return to this feature and discuss two possible

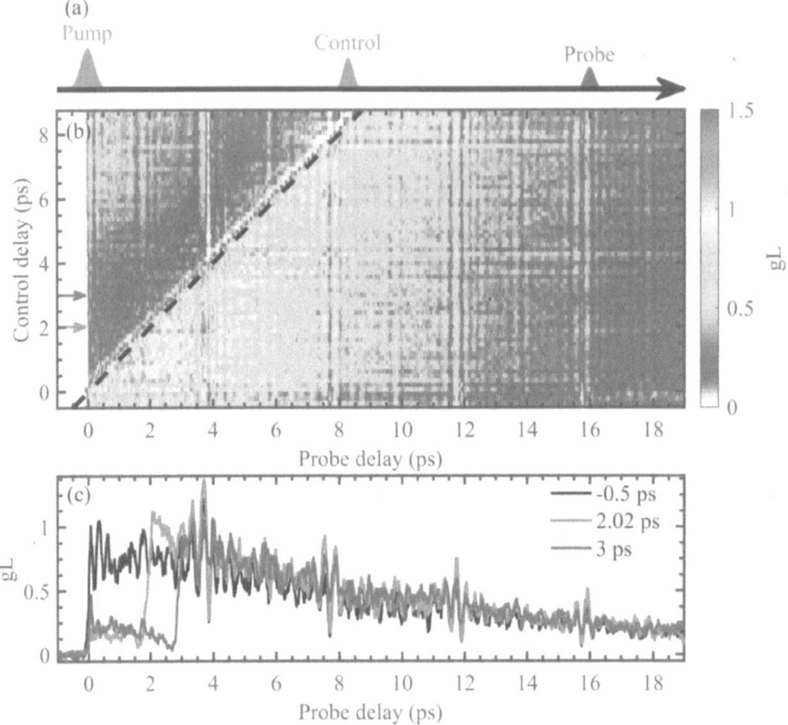

Figure 6.1: (a) The three pulses in the experiment. (b) gL (colour) as a function of the delay of both pulses. A superimposed diagonal dashed black line indicates where the probe and control delays are equal. Two coloured solid right arrows indicate the delays of lineouts in the bottom panel. (c) Horizontal lineouts from the above panel. $I_{pump} = 2.5 \times 10^{14}$ W/cm^2, $I_{control} = 10^{13}$ W/cm^2

explanations that arise from population transfer between the $X^2\Sigma_g^+$ and $A^2\Pi_u$ states. Beforehand, Sec. 6.1.2 confirms this coupling in a more intuitive configuration. As discussed in Sec. 5.2, the exponential-like decay of gain is consistent with the decay of population inversion, so the control pulse must modify the populations and phases of these states. Impulsive alignment by the control pulse also modifies the transient gain, which Sec. 6.1.1 confronts first.

6.1.1 Rotational excitation

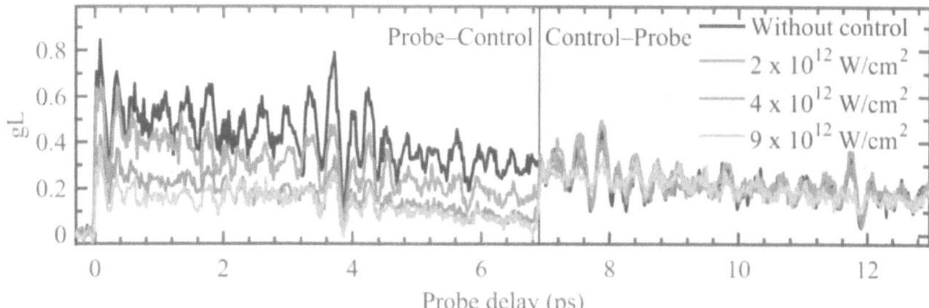

Figure 6.2: The order of the control and probe pulses divides the experiment into two delay regions (Probe–Control and Control–Probe). At a fixed control delay of 7 ps in this case, the results depend on the intensity of the control pulse. $I_{pump} = 1.9 \times 10^{14}$ W/cm^2.

Section 5.3.2 discussed the modulations superimposed on the decay of gain due to rotational wave packets in the $B^2\Sigma_u^+$ and $X^2\Sigma_g^+$ states. It showed that the $B^2\Sigma_u^+$ state rotational frequencies are stronger. The control pulse modifies the modulations in the Pump–Control–Probe regions of Fig. 6.1 and Fig. 6.2. For example, Fig. 6.1(c) shows that the amplitude of modulations is enhanced (2.02 ps) or reduced (3 ps) depending on the control delay. This behaviour is periodic and produces oscillations along the vertical direction of Fig. 6.1(b). At a control delay of 7 ps, the amplitude of modulations nearly disappears at control intensities near 10^{13} W/cm^2 in Fig. 6.2.

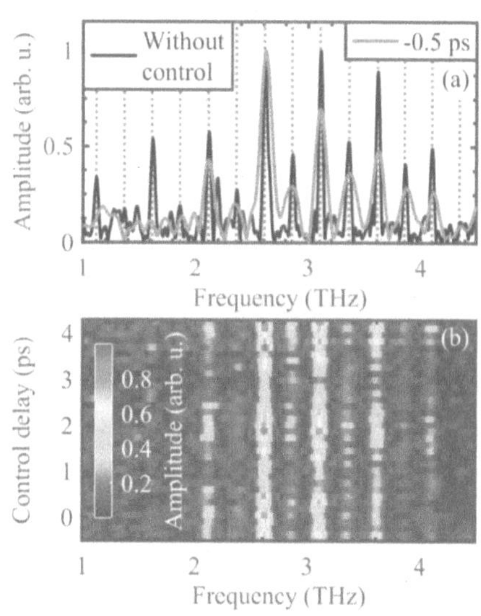

Figure 6.3: (a) Fourier transform of modulations with and without a control pulse at −0.5 ps. (b) The rotational frequencies oscillate as a function of control delay.

The Fourier transform of the modulations shows how the rotational wave packets change. Figure 6.3(a) shows the Fourier transform of the modulations at the negative control delay −0.5 ps, which includes data starting from a probe delay of 5 ps. For reference, the figure includes the Fourier transform without a control pulse from Fig. 5.6 and the expected frequencies of the dominant $B^2\Sigma_u^+$ state rotational wave packet. The $B^2\Sigma_u^+$ state rotational wave packet is also strong when the control pulse arrives before ionization. Figure 6.3(b) shows the Fourier transform at each control delay in Fig. 6.1(b) from a probe delay of 5 ps onwards. The amplitudes of the $B^2\Sigma_u^+$ state rotational frequencies oscillate with a period of about 2 ps as a function of control delay. The modulations oscillate with a similar period in the vertical direction of Fig. 6.1(b). This period is comparable to the quarter rotational revival time of the states.

The control pulse can remove the modulations by creating a new rotational wave packet that counteracts the initial rotational wave packet from the pump pulse [159, 160]. Increasing the intensity of the control pulse can confirm its role in exciting a new rotational wave packet. Figure 6.4 shows gain as a function of probe delay at two control delays using a higher than normal control intensity of greater than 3×10^{13} W/cm^2, which are compared to the case without a control pulse. At the control delay of 2 ps, the original modulations are nearly unchanged. At the control delay of 1 ps, new modulations emerge from the control pulse wave packet that are offset in time and overpower the original modulations. These results suggest that the control pulse creates new rotational wave packets in the ion that can complement and cancel the original rotational wave packets.

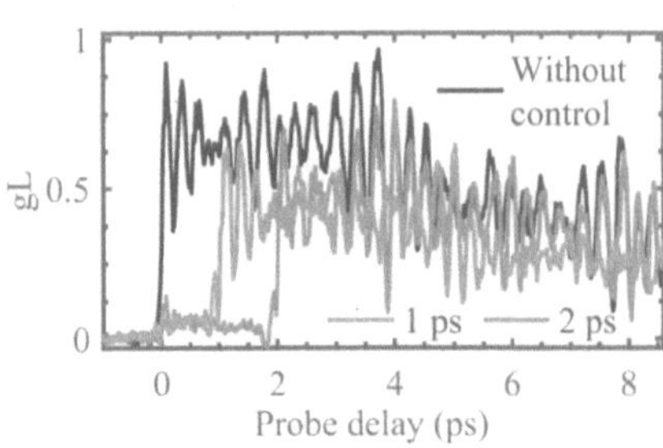

Figure 6.4: At a control delay of 1 ps, the rotational wave packets from the control pulse are visible. $I_{pump} = 2.5 \times 10^{14}$ W/cm^2, $I_{control} > 3 \times 10^{13}$ W/cm^2

6.1.2 Electronic and vibrational excitation

The control pulse can also influence the vibrational and electronic population distributions. Figure 6.5(a) shows the potential energy of the ground state of neutral nitrogen and the first three states of the cation. There are more than three vibronic transitions between the $X^2\Sigma_g^+$ and $A^2\Pi_u$ states that are within the bandwidth of the pump and control pulses [161], so both pulses move population between these states. This modifies the population difference between the $X^2\Sigma_g^+$ and $B^2\Sigma_u^+$ states and the gain measured by the probe pulse. This mechanism may be essential for pump-induced population inversion [61, 65, 85, 102, 103, 104, 105, 106, 107]. The control pulse should likewise influence the gain compared to the traditional measurement. This appears as a

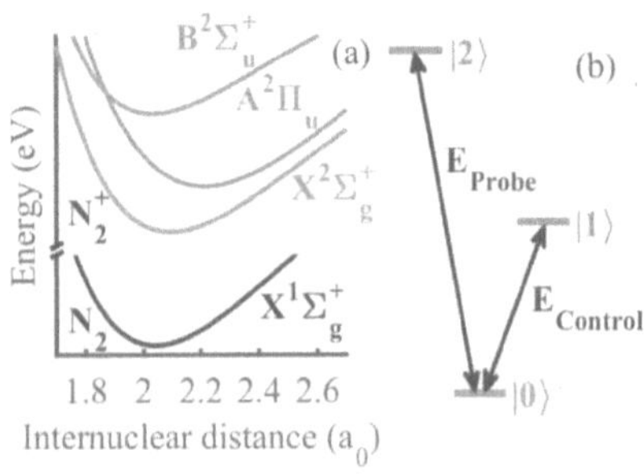

Figure 6.5: (a) Potential energy of the ground state of neutral N_2 and the $X^2\Sigma_g^+$, $A^2\Pi_u$, and $B^2\Sigma_u^+$ states of the ion. (b) Energy levels in the three-state V-system calculation.

different amplitude of the exponential-like decay in Pump–Control–Probe measurements. Section 5.2 attributed the decay to the decay of the population difference between the $X^2\Sigma_g^+$ and $B^2\Sigma_u^+$ states.

The amplitude of the decay increases due to the control pulse in Fig. 6.1(c), but this behaviour depends on the intensity and delay of the control pulse. For example, the amplitude of the decay decreases in Fig. 6.4 and is relatively unchanged in Fig. 6.2. Figure 6.6 shows gain as a function of control intensity with both delays fixed in the conditions of Fig. 6.2. It shows that gain at three probe delays generally increases as a function of control intensity up to about 1.6×10^{13} W/cm^2, but the two later probe delays also dip near 0.8×10^{13} W/cm^2. Rotational excitation may contribute to the dip, as the two delays overlap modulations in Fig. 6.2 that are reduced at a comparable intensity. This trend reverses at high control intensity, which may correspond to plasma defocusing from ionization of neutral nitrogen by the control pulse. These results show that the control pulse increases gain at non-ionizing intensities, but the change is too small to see in Fig. 6.2.

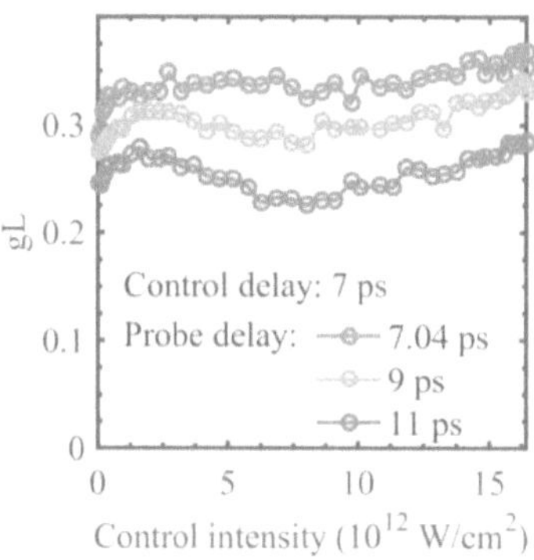

Figure 6.6: Gain as a function of control intensity at three fixed probe delays of 7.04, 9, and 11 ps after the fixed control delay of 7 ps. $I_{pump} = 1.9 \times 10^{14}$ W/cm^2

Figure 6.7 shows the relative control-induced change of gain

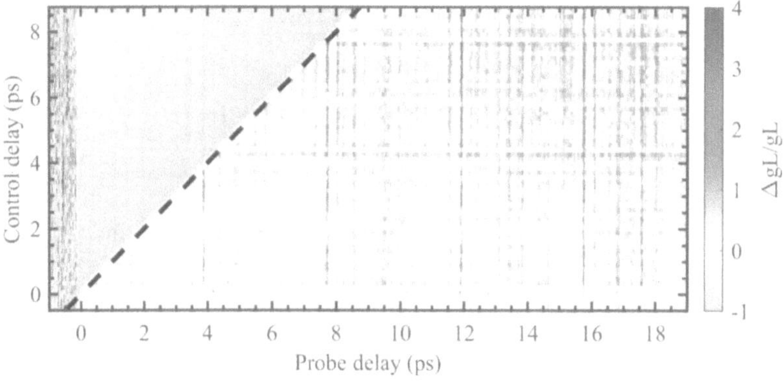

Figure 6.7: The control-induced change of gain ΔgL/gL from Fig. 6.1(b).

$$\Delta gL/gL = \frac{g_C L - g_0 L}{g_0 L} \tag{6.1}$$

that increases the visibility of the small changes to gain at long delays in Figure 6.1(b), where $g_C L$ is the measurement and $g_0 L$ is a reference measurement of gain without the control pulse. When the control pulse creates a complementary rotational wave packet that cancels the modulations, the peak and trough values are less extreme. This reflects a change to the transient gain and absorption in the system from rotational wave packets [89, 112] but is not necessarily associated with population exchange between the $X^2\Sigma_g^+$ and $A^2\Pi_u$ states. It is prominent in Fig. 6.7 and creates large positive peaks with 4 ps periodicity as a function of probe delay that oscillate as a function of control delay.

The Pump–Control–Probe region below the diagonal line of Fig. 6.7 is positive on average, and the negative regions are mostly consistent with transient gain modulations and experimental noise. Figure 6.7 and 6.6 show a control-induced increase of gain for a range of intensities and for all positive control delays, which is consistent with population transfer from the $X^2\Sigma_g^+$ to $A^2\Pi_u$ state.

6.1.3 Emission modification

Sections 6.1.1 and 6.1.2 showed the control pulse initiating dynamics that are measured by the probe pulse. In the Pump–Probe–Control configuration, the control pules arrives after the probe pulse and during the ongoing measurement. Section 2.3.1 showed that absorption and gain from electronic transitions can generate a free induction decay that trails the pulse while the coherence decays. In N_2^+, the duration of this lasing emission is tens of picoseconds long [63, 66, 67, 69, 79, 80, 85, 98, 100, 101]. The control pulse can modify the free induction decay, but the spectrometer integrates over the emission without time-resolution. The resulting gain is lower in Fig. 6.1, 6.2, and 6.4, so the control pulse must quench or redirect the free induction decay.

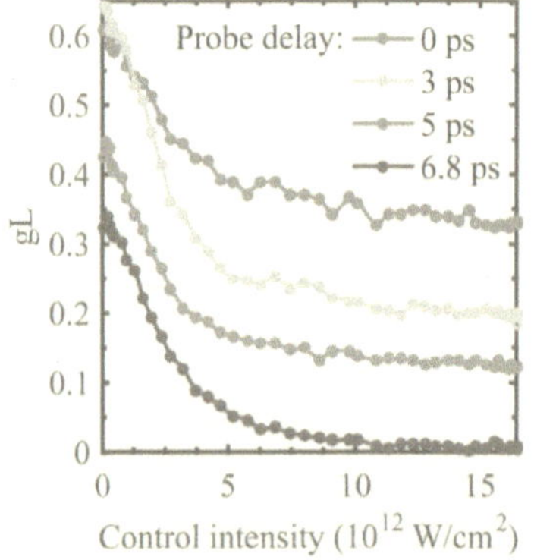

Figure 6.8: Gain as a function of control intensity at four fixed probe delays of 0, 3, 5, and 6.8 ps before the fixed control delay of 7 ps. $I_{pump} = 1.9 \times 10^{14}$ W/cm^2

Figure 6.8 shows gain as a function of control intensity at four fixed probe delays with the control pulse arriving later. The emission is modified at relatively low control intensities that also modified the exponential-like decay of gain in Fig. 6.6. At each delay, gain reaches a plateau at high control intensity.

The plateau contains the portion of the emission before the control pulse arrives, and it is correspondingly higher when the control pulse arrives later in the emission. This provides a measurement of the emission. The control pulse modifies the emission at all Pump–Probe–Control delays in Figure 6.7, but the modification stops when the control pulse follows the probe pulse by about 20 ps. This is consistent with the typical emission duration.

The emission could be halted or quenched, as both cases would prevent it from reaching the detector. Halting the emission requires quenching the coherence between the $B^2\Sigma_u^+$ and $X^2\Sigma_g^+$ states that was established by the probe pulse. Redirecting the emission requires changing the wave front of the emission. For example, a nonresonant pulse can induce a dynamic Stark effect [162, 163] and thereby redirect XUV free induction decay in attosecond transient absorption experiments [164]. In our case, the 800 nm control pulse is resonant with $X^2\Sigma_g^+$ to $A^2\Pi_u$ transitions and control-induced population transfer in this V-system can also modify the emission through the shared ground state.

6.2 Three level V-system

The $X^2\Sigma_g^+$, $A^2\Pi_u$, and $B^2\Sigma_u^+$ states form a V-system because transitions between the ground state and either excited state are allowed, but transitions between the two excited states are forbidden due to the parity selection rule. Furthermore, the transition at 391 nm [$B^2\Sigma_u^+$ ($\nu = 0$) $\rightarrow$ $X^2\Sigma_g^+$ ($\nu = 0$)] forms a V-system with the transition at 785 nm [$A^2\Pi_u$ ($\nu = 2$) $\rightarrow$ $X^2\Sigma_g^+$ ($\nu = 0$)]. The broadband 800 nm control pulse can influence the emission at 391 nm through the shared ground state of this system. Figure 6.5(b) shows a V-system, where the states are $|0\rangle$, $|1\rangle$, and $|2\rangle$. A calculation of this simple system can capture the essential physics by initiating emission on one transition and modifying it using the other transition.

The Hamiltonian for this system can be expressed in matrix form as $H = H_0 + V(t)$, where

$$H_0 = \begin{pmatrix} 0 & 0 & 0 \\ 0 & \epsilon_1 & 0 \\ 0 & 0 & \epsilon_2 \end{pmatrix}, \tag{6.2}$$

$$V(t) = \begin{pmatrix} 0 & -\mu_{10}E_C(t) & -\mu_{20}E_P(t) \\ -\mu_{10}E_C(t) & 0 & 0 \\ -\mu_{20}E_P(t) & 0 & 0 \end{pmatrix}, \tag{6.3}$$

ϵ_1 and ϵ_2 are the state energies, μ_{10} and μ_{20} are the transition dipole moments, $E_P(t)$ is the probe field, and $E_C(t)$ is the control field. The density matrix

$$\rho = \begin{pmatrix} \rho_{00} & \rho_{01} & \rho_{02} \\ \rho_{10} & \rho_{11} & \rho_{12} \\ \rho_{20} & \rho_{21} & \rho_{22} \end{pmatrix} \tag{6.4}$$

evolves according to the Liouville–von Neumann equation

$$\begin{aligned} i\hbar\dot{\rho} &= [H,\rho] \\ &= H\rho - \rho H. \end{aligned} \tag{6.5}$$

It provides the equations of motion for the density matrix elements, which form a coupled set of first order differential equations in atomic units

$$\begin{aligned}
\dot{\rho}_{00} &= -i\left(V_{01}\rho_{10} + V_{02}\rho_{20} - V_{10}\rho_{01} - V_{20}\rho_{02}\right) + \rho_{11}/T_{1,11} + \rho_{22}/T_{1,22} \\
\dot{\rho}_{11} &= -i\left(V_{10}\rho_{01} - V_{01}\rho_{10}\right) - \rho_{11}/T_{1,11} \\
\dot{\rho}_{22} &= -i\left(V_{20}\rho_{02} - V_{02}\rho_{20}\right) - \rho_{22}/T_{1,22} \\
\dot{\rho}_{01} &= -i\left(-\rho_{01}\epsilon_1 + V_{01}\rho_{11} + V_{02}\rho_{21} - V_{01}\rho_{00}\right) - \rho_{01}/T_{2,10} \\
\dot{\rho}_{10} &= -i\left(\rho_{10}\epsilon_1 - V_{10}\rho_{11} - V_{20}\rho_{12} + V_{10}\rho_{00}\right) - \rho_{10}/T_{2,10} \\
\dot{\rho}_{02} &= -i\left(-\rho_{02}\epsilon_2 + V_{02}\rho_{22} + V_{01}\rho_{12} - V_{02}\rho_{00}\right) - \rho_{02}/T_{2,20} \\
\dot{\rho}_{20} &= -i\left(\rho_{20}\epsilon_2 - V_{20}\rho_{22} - V_{10}\rho_{21} + V_{20}\rho_{00}\right) - \rho_{20}/T_{2,20} \\
\dot{\rho}_{12} &= -i\left(-\rho_{12}\left(\epsilon_2 - \epsilon_1\right) + V_{10}\rho_{02} - V_{02}\rho_{10}\right) - \rho_{12}/T_{2,21} \\
\dot{\rho}_{21} &= -i\left(\rho_{21}\left(\epsilon_2 - \epsilon_1\right) - V_{01}\rho_{20} + V_{20}\rho_{01}\right) - \rho_{21}/T_{2,21},
\end{aligned} \tag{6.6}$$

where V_{nm} are the nonzero elements of Eq. 6.3.

The equations of motion include phenomenological population and coherence decay times. The population decay times are $T_{1,11}$ and $T_{1,22}$ for $|1\rangle$ and $|2\rangle$, respectively. The coherence times are $T_{2,10}$, $T_{2,20}$, $T_{2,21}$; The first two correspond to the coherence time for the allowed transitions that are resonant with the control and probe pulses, respectively. Coherence between the two upper states is a by-product of the other two coherences, so its coherence time $T_{2,21}$ is the lesser of $T_{2,10}$ and $T_{2,20}$. The timescale for population decay is much longer than any of the dynamics that are investigated. The coherence times are more important because they determine the linewidth and the duration of the free induction decay.

The equations of motion can be solved using different methods. The Crank Nicolson method in time provides an accurate solution [165]. Applying this method, the first line of Eq. 6.6

$$\begin{aligned}
\frac{\rho_{00}^{n+1} - \rho_{00}^{n}}{\Delta t} = \frac{1}{2}[&-i(V_{01}^{n+1}\rho_{10}^{n+1} + V_{02}^{n+1}\rho_{20}^{n+1} - V_{10}^{n+1}\rho_{01}^{n+1} - V_{20}^{n+1}\rho_{02}^{n+1}) \\
&+ \rho_{11}^{n+1}/T_{1,11} + \rho_{22}^{n+1}/T_{1,22} - i(V_{01}^{n}\rho_{10}^{n} + V_{02}^{n}\rho_{20}^{n} - V_{10}^{n}\rho_{01}^{n} \\
&- V_{20}^{n}\rho_{02}^{n}) + \rho_{11}^{n}/T_{1,11} + \rho_{22}^{n}/T_{1,22}]
\end{aligned} \tag{6.7}$$

is approximated using the current value ρ_{00}^{n} at time step n and the unknown value ρ_{00}^{n+1} at the next time step. This can be rearranged, and the result

$$
\begin{aligned}
&\frac{2}{\Delta t}\rho_{00}^{n+1} - \rho_{11}^{n+1}/T_{1,11} - \rho_{22}^{n+1}/T_{1,22} + iV_{01}^{n+1}\rho_{10}^{n+1} + iV_{02}^{n+1}\rho_{20}^{n+1} - iV_{10}^{n+1}\rho_{01}^{n+1} \\
&- iV_{20}^{n+1}\rho_{02}^{n+1} = -i(V_{01}^{n}\rho_{10}^{n} + V_{02}^{n}\rho_{20}^{n} - V_{10}^{n}\rho_{01}^{n} - V_{20}^{n}\rho_{02}^{n}) + \rho_{11}^{n}/T_{1,11} \\
&+ \rho_{22}^{n}/T_{1,22}) + \frac{2}{\Delta t}\rho_{00}^{n}
\end{aligned}
\tag{6.8}
$$

includes only known values on the right-hand side. The left-hand side of Eq. 6.8 is a sum of products of known coefficients and the density matrix elements after the time step. Applying this procedure to every line of Eq. 6.6 creates a linear system $\vec{A} \cdot \vec{x} = \vec{b}$, where the matrix $\vec{A}$ is formed from the coefficients of the next density matrix elements on the left and the vector $\vec{b}$ is formed from the constants on the right. This system can be inverted to obtain the vector $\vec{x}$, which contains the density matrix elements at the next time step. Appendix A lists the full linear system.

Figure 6.9(a) shows the population of the three states as a function of time assuming control and probe pulse shapes composed of a transform-limited Gaussian function multiplied by a cosine function. The duration of the Gaussian pulse is 100 fs, and the fields are initially zero past the first cosine zero-crossing at 400 fs. The intensity of the probe pulse $I_{P,0} = 10^8$ W/cm^2 is relatively low, so it only induces a small part of a Rabi oscillation at 0 ps. The intensity of the control pulse $I_{C,0} = 4 \times 10^{12}$ W/cm^2 is high, so it induces several oscillations at 0.7 ps. The state energies are $\epsilon_1 = 1.54$ eV and $\epsilon_2 = 3.17$ eV, and the dipole moments are $\mu_{01} = 0.26$ a.u. and $\mu_{02} = 0.74$ a.u., which are similar to the parameters of the N_2^+ V-system. The center frequencies of both pulses are detuned by a couple of nanometers from each transition, which is also similar to the experimental case.

The initial populations in Fig. 6.9(a) correspond to a large inversion between $|0\rangle$ and $|2\rangle$ and no inversion between $|0\rangle$ and $|1\rangle$, which may be the case in N_2^+ after ionization [illegible]. The discussion also applies to free induction decay from absorption, as absorption and amplification only differ by a π phase shift. The coherence time $T_{2,20} = 5$ ps for $|0\rangle$ and $|2\rangle$ is a typical duration of N_2^+ lasing emission. In Fig. 6.9, the coherence time $T_{2,10} = 5$ ps of $|0\rangle$ and $|1\rangle$ is the same duration. The population decay times $T_{1,11} = 5$ ns and $T_{1,22} = 5$ ns are arbitrarily long.

The first-order wave equation in the reference frame of the probe pulse

$$
\frac{\partial E_P(t,z)}{\partial z} = -p_1 \frac{\partial E_P(t,z)}{\partial t} - p_2 \frac{\partial P_\mu(t,z)}{\partial t}
\tag{6.9}
$$

connects the polarization due to electronic transitions

$$
\begin{aligned}
P_\mu &= N\,\mathrm{Tr}(\rho\mu) \\
&= N\mu_{20}(\rho_{02} + \rho_{20})
\end{aligned}
\tag{6.10}
$$

back to the probe pulse that induced it, where

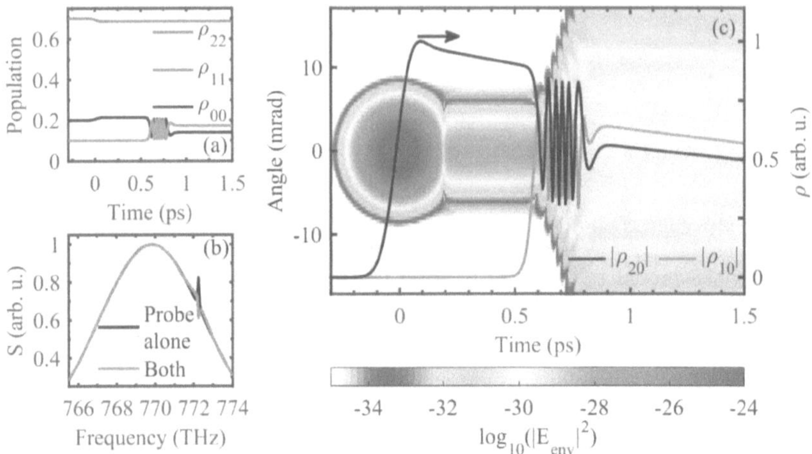

Figure 6.9: (a) The population of the three states as a function of time relative to the probe pulse. The total population was conserved to within less than 10^{-11} after 775,000 time steps. (b) The spectrum of the probe pulse with ("Both") and without the control pulse. (c) The spatial profile of the probe pulse as a function of angle and time on a logarithmic colour scale. The magnitudes of the off-diagonal matrix elements are superimposed.

$$\begin{aligned} p_1 &= \frac{1}{c} - \frac{1}{v_g} \\ p_2 &= \frac{2\pi}{c}. \end{aligned} \tag{6.11}$$

The wave equation for the control pulse likewise accounts for the polarization due to ρ_{01}, which we treat separately due to the opposite directions of the two transitions in the N_2^+ V-system. The probe field after one propagation step

$$E_P(t, z + \Delta z) = E_P(t, z) + \Delta z \frac{\partial E_P(t, z)}{\partial z} + \frac{\Delta z^2}{2} \cdot \frac{\partial^2 E_P(t, z)}{\partial z^2} + O(\Delta z^3) \tag{6.12}$$

can be expanded as a function of the step size Δz. Equation 6.9 provides the first-order derivative in Eq. 6.12. Differentiating Eq. 6.9 with respect to z provides the second order term

$$\frac{\partial^2 E_P(t, z)}{\partial z^2} = p_1^2 \frac{\partial^2 E_P(t, z)}{\partial t^2} + p_1 p_2 \frac{\partial^2 P_\mu(t, z)}{\partial t^2} - p_2 \frac{\partial^2 P_\mu(t, z)}{\partial z \partial t}. \tag{6.13}$$

The Euler or midpoint method can numerically approximate the derivatives in Eq. 6.9 and 6.13.

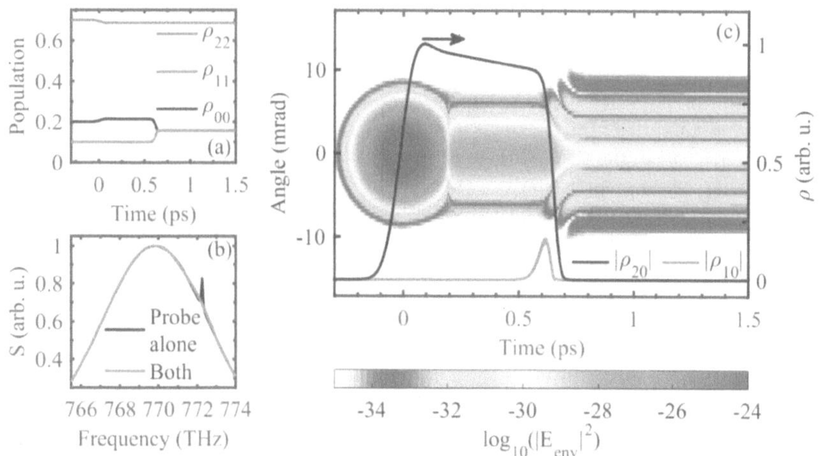

Figure 6.10: (a), (b), and (c) are identical to Fig. 6.9 except that $T_{2,10} = 5$ fs.

One step of propagation produces a free induction decay on the probe pulse. The spectra in Figure 6.9(b) are the Fourier transforms of the probe pulse after one propagation step with and without the control pulse. The control pulse modifies the lineshape, which corresponds to a phase shift in the emission and coherence [166, 167, 168]. The transverse spatial profile of the control pulse will induce a different phase shift at different positions in the beam, so the control pulse modifies the wave front of the free induction decay like a spatial light modulator [169, 170]. A detector in the far-field measures an electric field

$$E(k_x, k_y, \infty, t) = \iint_{-\infty}^{+\infty} E(x, y, 0, t) e^{i(k_x x + k_y y)} dx dy \tag{6.14}$$

that is related to the near field by a Fourier transform, which reduces to the Hankel transform of order zero for the simple case of cylindrical symmetry [11]. Numerically, non-uniform spacing along the radial axis enables a fast implementation of the Hankel transform [171]. The Hilbert transform provides the electric field envelope E_{env} to improve the appearance of figures.

Figure 6.9(c) shows the spatial profile of the probe pulse on a logarithmic colour scale as a function of angle in the far-field and time assuming that the probe and control pulses are Gaussian beams with waists of 50 and 75 μm, respectively. For reference, Fig. 2.3(c) showed the equivalent spatial profile without the control pulse. The control pulse significantly modifies the divergence of the free induction decay, which directs the remaining emission away from the propagation axis. This would reduce gain at an on-axis detector.

These three levels obviously do not capture all of the dynamics in N_2^+. Figure 6.5(a) shows that the equilibrium internuclear separation in the $A^2\Pi_u$ state is larger than the $X^2\Sigma_g^+$ state. Population transfer to the $A^2\Pi_u$ state on multiple transitions will launch a vibrational wave packet that can temporarily trap population [61, 102, 104]. The control pulse duration is longer than the timescale of vibrational dynamics, which may create a complicated mixture of the phases as population moves between the $X^2\Sigma_g^+$ and $A^2\Pi_u$ states and into and out of the trap. If these or other dynamics shorten the coherence time of $|0\rangle$ and $|1\rangle$, Fig. 6.10(a) shows that Rabi oscillations are heavily-damped and the populations of $|0\rangle$ to $|1\rangle$ rapidly equalize when the control pulse turns on. In this case, the coherence time $T_{2,10} = 5$ fs is very short, which destroys the peak in Fig. 6.10(b) and the coherence in Fig. 6.10(c). The wave front of the emission is still modified, but its amplitude is quenched.

The shared ground state allows the control pulse to modify coherence established by the probe pulse regardless of the coherence time. Pump–Probe–Control measurements are consistent with some combination of deflection and quenching. In both of these cases, gain is modified abruptly during the emission. The abrupt change of gain creates an abrupt change in the amplitude of the emission, which produces oscillations in the spectrum. Figure 6.11 highlights these small oscillations using the difference between the two spectra in Fig. 6.10(b). The period of the oscillations depends inversely on the delay between the pump and control pulses, which is similar to spectral interference between the free induction decay and the control-induced change to it.

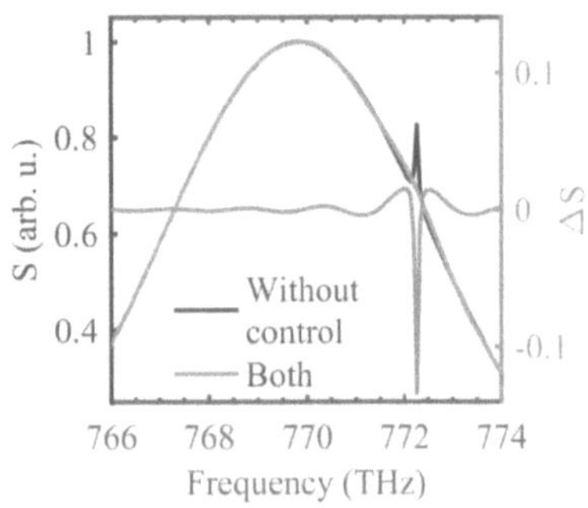

Figure 6.11: The spectra with and without the control pulse from Fig. 6.10 and the difference between them on a separate axis.

6.2.1 Propagation

The logarithmic colour scale in Figure 6.9 and 6.10 illustrates the important effects using only one propagation step and three states. Reality is more complicated. On subsequent propagation steps, the field after the first step in Eq. 6.13 becomes the input field in Eq. 6.3. The free induction decay itself generates further coherence as it grows during propagation. In N_2^+, beating between many transitions involving different rotational states creates a more complex free induction decay before propagation [69, 86, 98]. During propagation, the free induction decay experiences dynamic gain like Fig. 5.1 that amplifies it with additional modulation. In addition, it propagates in the presence of a refractive index that varies with time due to the molecular modulator discussed in Sec. 5.3.3. Therefore, propagation of the lasing emission is an important consideration.

6.2.2 Absence of rotational wave packet in the $X^2\Sigma_g^+$ state

Section 5.3.2 and 6.1.1 analyzed the rotational frequencies that modulate gain. The pump and control pulses both excite a stronger rotational wave packet in the $B^2\Sigma_u^+$ state than the $X^2\Sigma_g^+$ state. The weak $X^2\Sigma_g^+$ state rotational frequencies could indicate a large population inversion, but the V-system illustrates another possibility.

If $|0\rangle$ and $|2\rangle$ represent different rotational states in the $X^2\Sigma_g^+$ state, coherence between $|0\rangle$ and $|2\rangle$ is a rotational wave packet. A pump or control pulse at 800 nm could initiate the wave packet impulsively and simultaneously couple $|0\rangle$ and $|1\rangle$. The calculation shows that the pulses can modify or quench the rotational coherence as it forms between $|0\rangle$ and $|2\rangle$ by mixing $|0\rangle$ and $|1\rangle$, which may disrupt the formation of a coherent rotational wave packet. This system is distinct from the V-system because selection rules would also allow coupling of $|2\rangle$ and $|1\rangle$, but this only offers another path to modify the coherence between $|0\rangle$ and $|2\rangle$. Simultaneous $X^2\Sigma_g^+$ and $A^2\Pi_u$ coupling may disrupt impulsive alignment in the $X^2\Sigma_g^+$ state.

6.3 Recovering short-term gain

Section 4.6 and 5.4 investigate short-term gain that appears at low pump intensity near time overlap between the pump and probe pulses. It is consistent with Raman gain in V-system like Fig. 6.5(b) when the population of $|2\rangle$ exceeds $|1\rangle$. This process should also occur when the control and probe pulses overlap.

Figure 6.12(a) shows short-term gain at 428 nm in colour as a function of probe and control delay. When the control pulse arrives before ionization, the brief spike of gain near time overlap looks similar to Fig. 5.9. Short-term gain appears twice at positive control delays, which is consistent with wave mixing of the probe pulse with both the pump and control pulses. This process is modulated as a function of delay, which is clearer in Figure 6.12(b) at a fixed relative delay between the control and probe pulses. The 391 nm transition is still sensitive to molecular alignment relative to the laser polarization during Raman scattering, and the modulations are correspondingly similar to the alignment modulations superimposed on the exponential-like decay. At the control delay of 8 ps, the recovered gain in Figure 6.13(a) is sensitive to the polarization of the control pulse. This further confirms the influence from molecular alignment.

The control pulse reduces the initial short-term gain near zero probe delay in Fig. 6.12(a) and Figure 6.13. Raman gain should still induce a long free induction decay, so this behaviour is consistent with the quenching or redirection discussed in Sec. 6.1.3. Figure 6.13(b) further demonstrates this at a fixed control delay of 7 ps and two fixed probe delays that overlap with either the pump pulse (0 ps) or control pulse (7.04 ps). When the control pulse arrives 7 ps after the probe pulse at 0 ps, gain reaches a plateau like Fig. 6.8 because the control pulse can only modify the emission that follows it. At the probe delay of 7 ps, the recovered gain at 7.04 ps increases monotonically with control intensity.

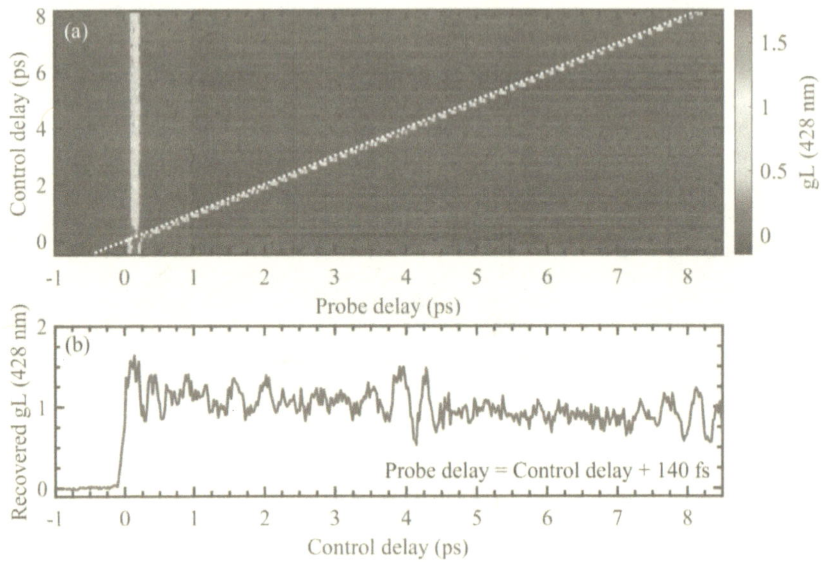

Figure 6.12: (a) Short-term gain at 428 nm in colour as a function of control and probe delay. The diagonal line indicates time overlap between the control and probe pulses. (b) Gain as a function of control delay at a fixed difference between the control and probe delay of 140 fs.

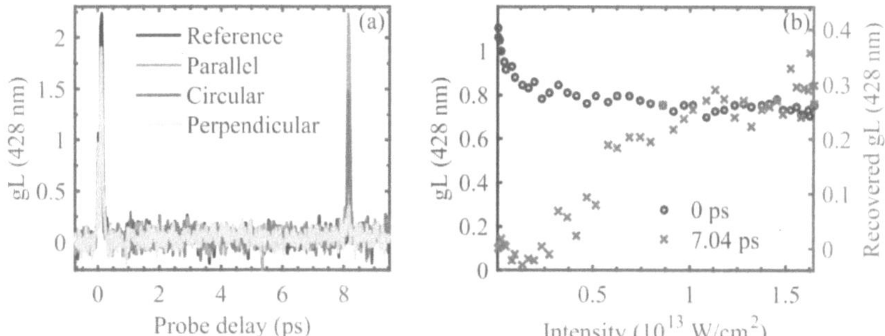

Figure 6.13: (a) Short-term gain at 428 nm as a function of probe delay using the control pulse polarizations indicated in the legend at a fixed control delay of 8 ps. (b) Short-term gain as a function of control intensity at time overlap (0 ps) and recovered at a later delay (7.04 ps). The control delay is 7 ps.

Finally, the control pulse can recover short-term gain, which further suggests the presence of Raman gain in the V-system.

Chapter 7

Conclusion

Gain in N_2^+ was first reported by focusing a laser tightly in ambient air and generating plasma. When a powerful laser is focused into a long gas medium, the experimental conditions become uncontrolled and difficult to characterize due to nonlinear effects that influence the pulse. In our experiment, a narrow gas jet in vacuum limits propagation so nonlinear effects on the pump pulse are insignificant.

We measure an exponential-like decay of gain on the timescale of electron-ion collisions in the plasma. Rotational wave packets in the states of the ion modulate molecular alignment, which is imprinted on gain and influences the propagation of probe pulses. The inelastic collisions that mix population also incoherently mix the rotational wave packets in the ground and excited states, so the rotational wave packets decay with gain. An appropriately timed control pulse can annihilate the rotational wave packets, which leaves only the exponential-like decay.

Recollision may transfer some population from the $X^2\Sigma_g^+$ to the $B^2\Sigma_u^+$ states, but it is not an essential gain mechanism. Like many others, our results suggest significant coupling between the $X^2\Sigma_g^+$ and $A^2\Pi_u$ states around 800 nm that depletes the ground state. The control pulse exploited this coupling to modify the emission following the probe pulse. This coupling may also explain the weakness of the rotational wave packet in the $X^2\Sigma_g^+$ state.

Short-term gain is distinct from the exponential-like decay in some conditions. It is consistent with Raman gain in a V-system involving the amplification of the probe pulse on transitions between the $X^2\Sigma_g^+$ and $B^2\Sigma_u^+$ states with simultaneous absorption of the pump or control pulse on transitions between the $X^2\Sigma_g^+$ and $A^2\Pi_u$ states. Short-term gain is hard to distinguish from the exponential-like decay without a probe pulse.

While we have not fully addressed the origin of the long-term gain, there are not many other apparent ingredients. It is possible that tunnel ionization and post-ionization coupling generate an inversion, which is strongly supported in literature by the experimental evidence of the coherence between the $X^2\Sigma_g^+$ and $A^2\Pi_u$ states and the enhancement of gain using polarization-shaped pulses. This field is still missing a critical experiment using an infrared pump pulse that does not couple the $X^2\Sigma_g^+$ and $A^2\Pi_u$ states after ionization. Similar to Ch. 6, a second 800 nm control pulse would then solely initiate dynamics in the $A^2\Pi_u$ state and definitively isolate the role of this seemingly critical component.

There is one missing component that relates to this book and may be important. Inelastic scattering of electrons with the $X^2\Sigma_g^+$ and $B^2\Sigma_u^+$ states taught us about the decay of gain, the decay of rotational wave packets, and the role of recollision. We did not address inelastic scattering with the $X^2\Sigma_g^+$ and $A^2\Pi_u$ states, and this cross section is not as well known.

In addition, we also did not address the role of superradiance. Two characteristics of N_2^+ lasing that are often attributed to superradiance are the delayed peak of the emission and the quadratic scaling of gain with density. For example, this is quite different from Fig. 2.3, which shows an exponential free induction decay whose intensity scales linearly with the density.

Inelastic scattering that equalizes the $X^2\Sigma_g^+$ and $A^2\Pi_u$ states could increase the population inversion between the $X^2\Sigma_g^+$ and $B^2\Sigma_u^+$ states over several picoseconds, which would generate a free induction decay that peaks later. In addition, the rate of this process would scale linearly with density like Eq. 5.5. If both the number of ions and the inversion increase linearly with density, gain would scale quadratically with density. This may be a fruitful alternative explanation for the characteristics attributed to superradiance.

Furthermore, inelastic scattering that equalizes the $X^2\Sigma_g^+$ and $A^2\Pi_u$ states would incoherently mix the rotational wave packets in both states, which is the same process that occurs between the $X^2\Sigma_g^+$ and $B^2\Sigma_u^+$ states leading to their rotational decoherence. The rate for this process could be higher for the $X^2\Sigma_g^+$ and $A^2\Pi_u$ states due to their smaller energy separation, which would cause a faster rotational decoherence in the $X^2\Sigma_g^+$ state compared to the $B^2\Sigma_u^+$ state. This is consistent with the relatively weak contribution from the rotational wave packet in the $X^2\Sigma_g^+$ state

In conclusion, an air laser device could be useful for remote sensing in scientific and other applications. However, this is a topic that deserves your attention not only because of a hypothetical device, but also because it manifests out of the fundamental physics highlighted throughout this book. Understanding the air laser will increase our understanding of other simple systems.

The title of this book encapsulates its core message: simplify the problem. If propagation is not an essential component of gain, limit it to make clearer measurements. Inhomogeneous conditions blur the results. Once we understand the simplest experiments, start adding the complexity of the real environment.

Looking towards the future, significant challenges lie ahead for air lasing research. It is an active field with many directions of independent research happening around the world, but the key components seem to have been identified and modeled already. Most papers seem to be specialized and report on individual phenomena under unique or poorly defined conditions. The field seems to lack a cohesive direction. There may be at least two components to the solution.

First, we should target the interdependence of effects in well defined conditions. While this book offers some insight already, there is room to take it a step further. For example, by using a weakly ionizing pulse followed by an alignment pulse, it should be possible to control lasing without inversion in the absence of the exponential-like decay of gain.

Similarly, Raman gain could be available in the same experiment. Then, as the intensity of the alignment pulse increases, the population decay should appear. This one experiment could contain three types of gain. Such an experiment could conclusively demonstrate the relevant phenomena, but theoretical agreement over a wide input parameter range would be necessary to optimize and customize the laser.

Second, it may be time to work towards a specific set of output parameters for a desired device. An ambitious goal would be to design and deploy an air laser for the purpose of remote detection. For example, fully-standoff detection was demonstrated with backwards-propagating amplified spontaneous emission (ASE) from neutral nitrogen lasing at 337 nm during filamentation. It required an additional forward-travelling Stokes pulse, but the tunability of the Stokes pulse enabled tunable remote sensing by stimulated Raman scattering [172]. While neutral nitrogen seems like an unlikely contender for the gain medium of the air laser, this approach has not been applied to the most promising gain medium. By studying N_2^+ lasing in this context, we will be forced to think about how its individual components can be useful.

This may be challenging because backwards N_2^+ lasing in air has not been studied extensively and may be difficult to achieve [124]. Backwards lasing is necessary for most applications of air lasing, so this should also be a target for future research. For example, it may be possible to assist the buildup of backwards ASE using a focus with a time dependent position that also travels backwards, which would increase the effective length of the gain medium. One proposal uses a series of pulse pairs with decreasing temporal separation. The second pulse in the pair travels faster due to atmospheric dispersion, so the earlier pulse pairs overlap at longer distances to create the moving gain medium [173]. Alternatively, it may be possible to achieve this by combining a chirped ionizing pulse with a focusing element that possesses a large chromatic aberration like a zone plate.

Finally, there are significant technical challenges associated with delivering the air laser reliably via filamentation. Fortunately, filamentation is a larger field of its own with a sophisticated and evolving understanding of the role of essential input variables like pump parameters, focusing conditions, and atmospheric conditions. By leveraging this knowledge, we can push air lasing to farther distances.

www.ingramcontent.com/pod-product-compliance
Lightning Source LLC
LaVergne TN
LVHW041232150826
845673LV00008B/2370

9783384210920